Society 5.0

Technological Prospects and Social Applications Set

coordinated by
Bruno Salgues

Volume 1

Society 5.0

Industry of the Future, Technologies, Methods and Tools

Bruno Salgues

iSTE

WILEY

Foreword

Society 5.0 Revealed

Prometheus and Atlas were right, one in having taught technology to people, and the other, in having skillfully managed the cities of his territory, Atlantis. Zeus was wrong in condemning them, one to have his liver eaten out every day after it grew back again during the night, and the other, to hold the world on his shoulders.

One day, “Atlas shrugged” and the world was shaken. *Atlas shrugged* is the voluminous novel written by the famous philosopher Ayn Rand in 1957. There, she summarizes her thesis, telling the story of the American economic collapse, when exploited businessmen under the control of demagogic politicians decided to give up. However, as it happens in every good American novel, it concludes on an optimistic note, where the head of the striker technologists, John Galt, comes out from his secret valley with his partner, saying: “Let’s go rebuild the world”.

This book by Professor Bruno Salgues is almost as dense. It deals with the same subject: the society we are creating. However, it is not a novel. He tackles all the facets of technologies that shape the world: of course the industrial revolution 4.0, but also new marketing approaches as well as the evolution of business management. This book transforms the weak signals that we all subtly feel into a superb synthesis of the societal evolution that is currently taking place.

Twenty-five years ago, at a national conference devoted to passive electronic components, Professor Bruno Salgues, whom I had never seen, spoke up and made relevant observations to the presentation of an industrialist, who was nonetheless an expert in his field. Puzzled by his remarks, I went to see him at the end of the conference to propose a later meeting and continue with his analysis. His reply was: “First, I would like to see your company and your factories, before discussing in detail”. That was done, and since then, we have regularly been seeing each other and discussing.

I still remember our joint trip to China in 2005, to visit an electronic equipment show. While I was paying attention to the products competing against mine, that Chinese firms were starting to manufacture, he was picking up faint signals and explaining the future trends he could deduce from them. Impressive!

This book, *Society 5.0*, is a societal summary, as interesting for a Master’s degree student, who will be able to develop a personal vision, as it can be for an experienced manager, who will find sources of inspiration.

I wish you an enjoyable read, especially considering that the concrete examples told throughout these pages could be the tales of Greek mythology of the present-day.

Joseph PUZO[1]

1 CEO of AXON’ CABLE group, a member of the steering committee of the national-extent project known as Future Industry (*Industrie du futur)*. In 1980, AXON’ used to be a SME of 100 employees, in charge of standard cable manufacture. In 2018, it has become a transnational (a particular multinational organization) SME of 2,100 employees, designing and providing innovative cabling solutions for the space conquest of the Moon, Mars and beyond, for the super magnets that enabled the discovery of the Higgs boson, for surgery, for the flight controls of many planes, for environmental remediation and for car airbags.

Preface

Together with the one I wrote a few years ago on mobile phones[1], this book has been one of the most difficult ones to write and to conceive, because in this developing society 5.0, everything is interconnected. It foresees the birth of the "firms of the future", which are the topic of many speeches and white papers, many of which are not up to the mark. This type of society relies on a series of concepts: automation, dematerialization, digitization, industrialization and "servitization", which shake up economic and political life. These concepts are at the origin of new actors, the death of powerful and recognized organizations, while other organizations see how their mutation takes place, a mutation which is itself complex, brutal, but very real.

Society 5.0 involves an important social change. It results from the birth of technologies, which have become mature and have been released in record times, something that no other technology had done in the previous century.

Had Paul Valéry foreseen the arrival of society 5.0? He wrote:

> "What can result from this great debauchery [...] the amount of publications, their daily frequency, the flow of things which are printed and published, model judgment and impressions from morning to night, mix them up, and transform our brains into a truly gray substance, where nothing lasts, nothing dominates,

1 Bruno Salgues, *Les Télécoms mobiles*, Hermes Science, Paris, 1995.

> and we experience the strange impression of the monotony of novelty and the boredom of wonders and extremes".[2]

The theme of the informational deluge and infobesity, which imposes an adequate processing of information, will be a constant in this book. As a corollary and an opposing element, boredom will be one of the elements of the suggested managerial approach.

Let us quote Paul Valéry once more:

> "But the individual also means the freedom of the mind. Now, we have seen that this freedom (in its highest sense) becomes illusory by the mere effect of modern life. We are influenced, harassed, made stupid, vulnerable to all the contradictions, to all the dissonances that tear the environment of the present civilization. The individual is already compromised even before the State has fully assimilated him".[3]

I would like to thank those who had the courage to read, amend and criticize the first drafts of this book, including Alexis, Joseph, Monique-Marie and Philippe.

This book is not an anti-manual of politics, management or technology marketing, or a theory of innovation. It is a book that simultaneously wishes to fight against the excesses of this society 5.0 which is unveiling, a society that is often misunderstood, and to bet on introducing the ideas that make it up, as well as its associated technologies, all working for the improvement of society. This book may appear as a manual full of philosophy or definitions, and I do apologize for this beforehand to some readers, but that is the way it is.

Bruno SALGUES
June 2018

---

2 Translation of French quote. Quoted from Paul Valéry in Jean-Pierre Siméon, *La Poésie sauvera le monde*, p. 36, Le Passeur Éditeur, Paris, 2015.

3 Translation of French quote. Paul Valéry in "*Le bilan de l'intelligence*", extracted from a conference delivered in 1935, published by Allia Editions in 2011.

Introduction

Preamble for Understanding Society 5.0

In June 2017, Paul Reyntjens, financial policy officer for De Lork association, published on his LinkedIn profile a text in French which inspired this preamble[1]. The quotes for this introduction have been extracted and translated from there.

> "In 1998, Kodak had 170,000 employees and sold 85% of photo paper in the world. In a few years, their turnover collapsed and Kodak went bankrupt. What happened to Kodak will happen to many companies in the next 10 years and people do not see it coming.
>
> In 1998, would you have thought that 3 years later you would never be able to take pictures on film paper again?"

It should be noted that these empire disappearances were not the exclusive fate for the firms of that generation: AOL, Lycos, Yahoo! and ouiEurope did not have the same success, although for some time they were the "stars" of the new economy, which was the term used at the time of their launch. Photo cameras and digital cameras were invented in the 1970s, notably by Kodak. In their beginnings, they had a poor-quality resolution, according to photographers. They barely had 10,000 pixels, whereas now they have tens of millions. As usually happens with all new technologies, these devices were disappointing for a long time. They even seemed unusable. Some gurus were very clear and anticipated that these would not succeed. However,

1 Available at the following address: https://fr.linkedin.com/pulse/tres-interessant-%C3%A0-lire-le-monde-%C3%A9volue-quel-paul-reyntjens.

the reality is different. They first replaced conventional technologies, and worse, they became better! All these technologies are characterized by two "occult" forces, which are digitalization and dematerialization.

Classical economy, that of society 1.0, was based on agriculture.

> "In the future, there will be a farmer robot valued at $100. Farmers in the third world will be able to manage their fields, instead of working all day long. Hydroponics will require much less water. Veal meat produced in laboratories is already available and will become cheaper than natural veal by 2018. Currently, 30% of the entire farmland is used for livestock. Imagine if we did not need it anymore. Several new companies will soon launch insect proteins on the marketplace. These are richer than animal proteins. They will be labeled alternative sources of protein".

Agriculture, the foundation for society 2.0, is currently undergoing important changes, which will have to be studied to see its applications in society 5.0. This agrarian society was called into question by the industrial society, that of society 3.0.

The strength of industrial society lies in the power provided by forms of energy such as electricity, which revolutionized the industry. For example, with the arrival of electricity, the textile industry no longer needed to be placed near energy sources. These sources were coal in the Roubaix area in France, or Hesse in Germany, and electric dams in the southern foothills of the *Massif Central*. The location of industries depended on these sources. Since the topic is fashionable, many academics who have never worked at a factory have recently dealt with this subject with more or less success.

> "Electricity will become incredibly clean and inexpensive. Over the past 30 years, solar production has been growing exponentially. We are only beginning to see the impact. Last year, there was more energy produced by solar sources than from fossil sources, worldwide. The price of solar energy will decrease so much that every coal mine will stop being exploited by 2025.
>
> Cheap electricity means plenty of water at a low price. Desalination now only needs 2 kWh per cubic meter. In the

> majority of cases, it is not water which is uncommon, but it is drinking water that is. Imagine what could be possible if everyone could have unlimited drinking water for almost anything".

The emerging society will free itself from the problems associated with energy location and will be able to relocate its production sources thanks to the information society. Society 4.0, the information society, which came after the industrial society, is the basis for this new society that we are studying. Thus, the fourth industrial revolution will transform most of the traditional industries between the next 5 and 10 years, by means of information and communication technologies and knowledge-related tools. Among these technologies, we can mention artificial intelligence, robotics and 3D printing.

I.1. Artificial intelligence

Regarding artificial intelligence, computing tools, the possibilities of memories and algorithms are becoming more and more efficient, and this is happening exponentially. Besides, there are more and more effective methodologies for understanding the world. After having beaten chess players in the 1970s due to "logical errors" in programs, a computer finally beat the best player of Go in 2016. In artificial intelligence, errors are beneficial and in general, precision is symbolic.

> "In the United States, young lawyers find it difficult to be employed. This is because IBM's Watson computer can provide legal advice in a few seconds, for more or less complicated cases, all with 90% accuracy, compared to 70% for humans. So, if you are studying Law, forget it right away. In the future, there will be 90% fewer lawyers, only those who are specialized will survive.
>
> The Watson computer is already helping to diagnose cancer with 4 times more accuracy than humans".

The holder of this singularity believes that, in 2030, a computer will be able to compete against human intelligence. Thus far, face recognition software has proved superior to human capabilities, and it can be used for biometric access recognition, for searching people in applications like Facebook, and, as a matter of fact, has dethroned professional physiognomists.

I.2. 3D printing

3D printing makes it possible to print everything, from the smallest items – a few nanometers – to houses. This kind of technology plays an important role in the evolution of goods production.

> "In 10 years, the price of basic 3D printers has gone from $18,000 to $400. At the same time, they have become 100 times faster. All major shoe manufacturers have started to print shoes. In airports, spare parts are already printed in 3D. The space station has a printer which does not need to have as many spare parts as before. By the end of 2017, new smartphones will have been able to digitize in 3D. Thus, it is possible to digitize your feet and print perfectly adapted shoes at home.
>
> In China, a 3D building of 6 complete floors has already been printed. By 2027, 10% of everything that will be produced will be done by 3D printer".

I.3. Platform economy and "servitization"

The birth of the platform economy is a reality. Unfortunately, this economy has been the object of conspicuous ignorance on the part of theoreticians. This new economy is characterized by the passage from a possession-oriented economy to a new economy, featuring a form of "servitization", a phenomenon which we will describe in this book. This economy of the platform is wider than that of sharing.

> "Uber is simply a software tool. Even though it has no cars, Uber has become the largest taxi company in the world. Airbnb is currently the largest hotel chain in the world, despite the fact that it does not have any facilities".

1

Society 5.0, Its Logic and Its Construction

The main theme of this book is the description of society and of everything which surrounds it, that is to say, what has been defined as the fifth societal wave.

1.1. The origins of society 5.0

The term "society 5.0" first appeared in Japan in 2016. Since then, it has been spreading and its underlying concepts continue to be shaped.

Society 5.0 is a term used in the *Fifth Science and Technology Basic Plan*, reviewed by the Japanese Government's Council for Science, Technology, and Innovation. It was enforced by the Japanese Cabinet of Ministers in January 2016.

DEFINITION. "Society 5.0" can be defined as a "society of intelligence", in which physical space and cyberspace are strongly integrated.

Society 5.0 emerged from the hunter-gatherer society, the agricultural society, the industrial society and the information society. Although focused on humanity, 5.0 refers to a new type of society where innovation in science and technology occupies a prominent place, with the aim of balancing social and societal issues that need to be solved, while ensuring economic development. Although it borrows many of its elements, this approach is opposed to that of the proponents of decline.

The Japanese have introduced society 5.0, by illustrating it with Figure 1.1.

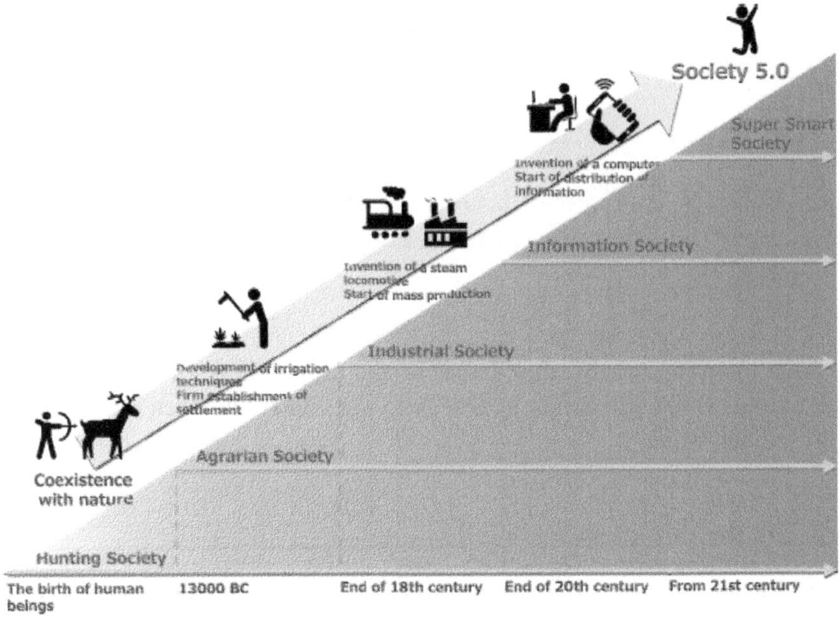

Figure 1.1. *Definition of society 5.0. For a color version of the figure, see www.iste.co.uk/salgues/society.zip*

The "basic plan for science and technology" is a comprehensive plan for the promotion of science and technology, decided by the Japanese Government and incorporated into a fundamental law, which will be positioned in the next decade, in the same way as the European H2020 bid solicitations. The fifth basic plan (2017–2021) is the first one for Japanese governments. This law establishes a Council for Science, Technology, and Innovation (CSTI). Thanks to an array of different measures, it strongly favors innovation policies for science and technology. The plan has been brought forward as a model for a wide range of stakeholders, including government, the university environment, the industry and citizens, so as to co-build the society of tomorrow. It also aims to guide Japan so that it can become “the most favorable country for innovation”.

This plan is sharply different from the innovative strategic axes proposed by the French Minister of Economy, Arnaud Montebourg, and amended by his successor, Emmanuel Macron. In Europe, countries have defined industrialization axes in general, and this was certainly the case in France. The Ministry for Productive Recovery in office in those times translated the priorities of the French industrial policy into 34 plans, which were designed for coordinating the actions of public actors and companies. The announced aim was to favor the emergence of the products of the future under the label “Made in France”.

– Big Data
– Cloud computing
– Connected objects
– Augmented reality
– Embedded software and embedded systems
– Contactless services
– E-education
– Digital Hospitals
– Cybersecurity
– Nanoelectronics
– Robotics
– Supercomputers
– Innovation in the food industry
– TGV of the future
– Two liters per 100 km vehicle for all
– Autopilot vehicles
– Electric recharging terminals
– Battery autonomy and power

– Electric planes and a new generation of aircrafts

– Electrically propelled satellites

– Heavy load airships

– Ecological ships

– Renewable energies

– Smart electrical networks

– Water quality and management of scarcity

– Medical biotechnologies

– Innovative medical devices

– Telecommunications sovereignty

– Wood industry

– Recycling and green materials

– Thermal renovation of buildings

– Green chemistry and biofuel

– Technical and intelligent textiles

– Factory of the future

Box 1.1. *The initial 34 industrial plans*

When his successor, Emmanuel Macron, took office, he reduced Montebourg's industrial plans from 34 to 10. The plan was associated with goals. These 10 plans were broken down into 9 specific plans, and 1 particular and transversal plan called "*Usine du future*" ("Factory of the future"). This approach was considered more important than others. The "Industry of the future" approach sought to develop a technological offer, by accompanying the transformation of companies, training employees, and with a strong emphasis on international cooperation.

Plan	Contents	Goals
New resources	Biosource and recycled materials	2020: to double the volume of raw material of vegetal origin in the French chemical industry, increase nondangerous waste recycling by 50%
Sustainable city	Water, smart grid, thermal renovation, wood industry	2020: 100 billion euros turnover and more than 110,000 territorialized jobs
Ecological mobility	2 L/100 vehicle, charging station, autonomous vehicle, energy storage	2016: 20,000 recharge stations[1] 2020: 30% decrease in CO_2 emissions of new vehicles manufactured in France
Transport of the future	TGV, ecological ships, electric planes, electric airships and drones	2020: to sell 80 electrically run school-planes, undated 50% decrease in energy consumption of buildings
Medicine of the future	Digital health, medical biotechnologies, medical devices	2017: 50,000 patients with a chronic disease under medical remote monitoring[2] 2025: to reestablish the French trade balance in terms of medical technology, whose current deficit is equivalent to 1 billion euros
Data economy	Big Data, supercomputer, Cloud Computing	2020: to create and consolidate 137,000 jobs thanks to Big Data, to master the critical technology of exascale supercomputers, which make it possible to perform 1 billion operations per second
Intelligent objects	Connected objects, robotics, augmented reality, contactless services, innovative textiles	2020: 8 million customers using mobile payment methods, to deploy an interoperable ticketing app to be used in 50% of cities with more than 200,000 inhabitants

1 According to data provided by the roaming consortium for electric mobility recharge GIREVE (*Groupement pour l'itinérance des recharges électriques de véhicules*), by the end of 2016, in France, there were 15,883 public parking places for recharging electric vehicles, disseminated in 5,297 stations. For more information: http://www.gireve.com.

2 If we only take heart failure into account, 350,000 pacemakers and 60,000 connectable automatic defibrillators are implanted in France. However, only 40,000 of these devices can be activated by a remote monitoring system.

Plan	Contents	Goals
Digital trust	Cybersecurity, telecommunications sovereignty, nanoelectronics, embedded systems, embedded software, electrically powered spacecraft propulsion	2020: to develop 5G infrastructure per mile, annual growth of 30% in the cybersecurity export market share, half of all sales in electric satellites
Intelligent food	Functional food, packaging of the future, lasting cold and food security	2017: modernization of 30% of industrial slaughterhouses and recruitment of 90,000 people for the sector

Table 1.1. *Emmanuel Macron's nine specific plans*

1.2. The ancient ages

Society 5.0 can be defined by its uses and by its societal leading elements.

After the 18th Century and the birth of the industrial revolution, innovations followed one another.

1771: the beginning of the Industrial Revolution with the appearance of machines, factories and canals.

1829: steam, coal, steel and train became the four predominant development factors.

1875: steel and hard engineering encouraged the birth of new development stars: electricity, chemistry, civil engineering and shipbuilding. Governments sponsored engineering schools in these fields.

1908: the car became complementary to water and railway means of transport. It quickly required oil resources, which led to the emergence of complementary industries, such as petrochemistry and mass production.

1971: the information and communication technology industry fed multiple economic sectors, leading not only to the birth of a semiconductor industry, but more fundamentally to the world of software and applications.

OBSERVATION. Society 5.0 is not defined by these innovation waves, but by the way in which innovations have modeled society. Society 5.0 is based on different pillars, including industry 4.0 and cybernics. It has changed lifestyles with the appearance of city 3.0. This society employs information and communication technologies 2.0 and defines humanity 2.0.

Each of these themes will be discussed in the rest of the book.

1.3. Cybernics or cyber-physical systems

DEFINITION. Cybernics is an interdisciplinary field that aims to develop technologies, industries and societies, which can help to support and examine the functions of the human body.

Cybernics includes various fields, such as the development and spread of medical robots and nursing care, engineering, medical sciences, information science and social sciences. These technologies are often designated as "cyber-physics".

Cybernetics was founded by Norbert Wiener at the times of World War II[3].

DEFINITION. Cybernetics is the science of governing or managing information, with the goal of driving systems.

Cybernics is the tool implemented for carrying out cybernetics. Cybernetics relies on:

– the notion of control;

– energy regulation;

– entropy reduction.

Norbert Wiener's approach relies on the notion of feedback, which translates a circular vision of communication. The notion of energy regulation is necessary in order to set up society 5.0. At Yoshiyuki Sankai, cybernics is introduced as a frontier science that combines cybernetics, mechatronics (itself an alliance between mechanics and electronics) and

3 We are referring to Norbert Wiener's book, *Cybernetics, or Control and Communication in the Animal and the Machine*, MIT Press, Boston, 1961.

computing. Its goal is to integrate humans into robotics and the other way around. One example of this could be the Hybrid Assistive Limbs (HAL) exoskeleton. For Kenji Suzuki, it should be praised as a technology which can lead to an augmented man.

1.4. The Council on Competitiveness-Nippon (COCN)

In order to increase industrial competitiveness, which is the foundation for the sustainable development of a country, various policies are implemented, particularly in terms of science and technology policies and industrial policies. The roles of the public and private sectors are coordinated by the Competitiveness Council. They issue common policy recommendations regarding the industry, cooperation between different universities, the government, its agencies and other actively related organizations. A group of 38 interested firms, present in various industrialized countries, are currently working in order to encourage and carry out the project.

1.5. The lessons of history

The contributions of societies from the past are summarized in Table 1.2.

Society	Characteristics	Society 5.0
1.0 Hunting and gathering	The need for sustainability	Full use of information and communication technologies
2.0 Agricultural	Inclusion	Citizen-centered (included citizens)
3.0 Industrial	Effectiveness	Participation of everyone (rejection of fractures)
4.0 Information	Power of intelligence (and knowledge)	Shared values: sustainability, inclusion, effectiveness and the power of intelligence

Table 1.2. *Characteristics of society 5.0*

Society 5.0 appears as the will for balance in the search for optimization of the four previous societies.

Society 5.0 seeks to achieve sustainability (ecology), broad inclusion, efficiency, and therefore, the industrial competitiveness of those who implement it using the power of intelligence and knowledge.

1.6. The decision variables of society 5.0

It is necessary to raise the question regarding the decision variables of society 5.0.

QUESTION. Are information, distractions, identity, alienation and action other decision variables?

What is interesting in our case is that this questioning leads back to our psychological chain and the notion of boredom. The following chapters will develop these topics.

1.6.1. *Which role for information?*

The flagship product of media is information, in fact; it is a pure product that eliminates boredom, because information is interesting. "Information is processed, meaning is interpreted"[4]. The romantic and modern vision has pushed men to integrate collective modes of thought, which quickly seem limited, and generate boredom. These ways of thinking are built by media and advertising. On the other hand, information is not ready to die away, so this causes infobesity.

For Ladwein, information is deployed on two levels: the informational context and that of basic knowledge. The implementation of memory, perceptions and interpretations is at the origin of reasoning, of judgment, and this leads to deliberation, choices, which are then translated into behavior[5]. This fundamental mechanism should not be rejected on the pretext of boredom and the chain "need-desire-expectation-demand".

Marketing experts like Nathalie Joulin have pointed out that consumers are better informed and more autonomous. Consumers are more experienced

4 Lars Fr. H. Svendsen, *Petite philosophie de l'ennui*, translated from Norwegian by Hervieu H., p. 41, Fayard, Paris, 2003.

5 Richard Ladwein, *Le Comportement du consommateur et de l'acheteur*, 2nd edition, p. 51, Economica, Paris, 2003.

and have been modeled by new technologies, media and traveling. This kind of knowledge has made consumers more independent. For example, we can observe that in the field of health, computers and their software will increasingly replace doctors[6]. This notion of information integration is crucial for society 5.0. We should observe that the notion of information integration into processes and actions is more important than artificial intelligence, as we know it these days.

The other point is the frequent distinction between code-information and signal-information. The so-called "information theory" of the mid-20th Century (Shannon, Wiener) employed the notion of signal-coding at the transmission end. Then, a statistical theory was superimposed on the initial theory. The idea of using Boolean algebra contributed to the notion of computationism. In fact, the concept of computationism went beyond the notion of computation, and gave rise to the fact that coded-information has a double, which is the signal, meaning that we are able to process not only calculations, but also events. Indeed, the latest developments in computer science have stemmed from this. New keywords, such as the Internet of Things (IoT), were born. Events surveyed by these objects will lead to actions.

DEFINITION. Computer science expresses the theoretical and technical possibility of manipulating signals, which correspond to numerical symbols, and can thus perform calculations.

DEFINITION. A computer is a set of signals managed by algorithms, with a pre-established purpose.

The information code generally has cognitive content, or meaning, and this is regardless of the type of media employed. The term "cognitive share of information" is also used. The signal information is an indication, or a stimulus. In this case, the corresponding notion can be represented by a concept.

The informational approach generated what is known as economy 2.0, where both types of information are required.

6 Nathalie Joulin, *Les Coulisses des nouveaux produits*, p. 18, Eyrolles, Paris, 2002.

1.6.2. *Which role for time?*

Boredom can only really take place, provided that everything, every action has its own temporality. Philosophers such as Heidegger made a distinction between being bored about something and getting bored while doing something. Killing time is not given in a situation, it is rather the situation itself that becomes the means for killing time. A state of alert is characterized by the fact that the notion of time is assimilable to the moment (*kairos*). This differs from clock time (*kronos*), which is defined as specific to the thing. On the other hand, *kairos* is related to *krisis*, the crisis, the decision, that is to say, to the information signal.

In all cases, time is full to such an extent that it disappears in sheer transparency, as Svendsen cleverly pointed out[7]. For him, it is precisely the temporality of daily life, which engenders indifference in the world, and consequently, boredom. Objects are brought closer to men in a strange absence of difference and these objects transmit code-information.

Consultants state that they have observed a new relationship towards time which transforms quantitative time into qualitative time. For them, time is the biggest innovator, in that it elapses and everything happens according to it. Consumers evolve in terms of tastes, mentality and habits[8].

The impression of an acceleration of time, that is to say, of acceleration in the deployment of *kronos*, had led some authors to suggest living more slowly, in a Slow Economy.

1.6.3. *Which role for nature?*

As Jean-Paul Demoule[9] observed, the invention of agriculture and livestock introduced a different relationship towards nature.

> "Hunter-gatherers feel immersed in nature. When an animal is about to be killed, the hunter asks for permission from the animal or the animal spirits. When we want to express our

---

7 Lars Fr. H. Svendsen, *op. cit.*, p. 182.

8 Nathalie Joulin, *op. cit.*, p. 63 and following pages.

9 We are referring to the book by Jean-Paul Demoule, *Les Dix Millénaires oubliés qui ont fait l'histoire*, Fayard, Paris, 2017.

world view, we also do it through animals, as we can see it in decorated caves. Becoming a breeder implies a radical reversal of this world view, it is like getting away from nature: hunter-gatherers had successfully domesticated the dog from the wolf, but this was rather a win-win type of combination, which was not the case in domestication for meat".[10]

1.6.4. *Which role for distraction?*

Distraction is one of the means for fighting boredom. It is also a source of activity in society 5.0.

From a historical point of view, this was the task of royal entertainers. Between 1620 and 1642, the Marquis de Cinq Mars became the entertainer of King Louis XIII, a difficult character to amuse. Having become Grand Squire of France, the Marquis was unhappy to live with a bored man. Indeed, distraction was set at the core of society 5.0 during its development. In Copenhagen, Tivoli was a forerunner in the field of leisure park construction.

Theme parks and tourist attractions are examples of this quest for distraction in society 5.0. This notion brings together a multitude of establishments and places of very different natures and sizes:

– animal parks, like safari-parks, which came after zoos;

– recreational parks, defined as enclosed spaces, devoted to play and entertainment. One of their key elements is the explicit display of a commercial vocation. These are less oriented towards outdoor recreation than a leisure center;

– aquatic-oriented parks, whose operating costs are close to those of a recreational sports pool;

– amusement areas, associated with leisure centers or tourist sites;

10 Translation of French quote. Quotation extracted from an interview to Jean-Paul Demoule in *Le Monde* newspaper, from October, 29th 2017. Available at the following address: http://www.lemonde.fr/sciences/article/2017/10/29/la-revolution-neolithique-n-a-pas-de-comparaison-dans-l-histoire_5207496_1650684.html# MhqLp4OZTHb SOQCu.99.

– cultural or educational parks, which employ attractions in their museography (robots, train, etc.). These facilities feature a pedagogical vocation, which rather assimilates them with ecomuseums, modern museography having increasingly integrated more playful and recreational techniques.

France offers its public more than 300 parks, but some of these are small and seasonal. Disneyland Paris, Grévin and Futuroscope capture the largest audience. This number greatly exceeds other European countries, which might distinguish France from the rest and be an indicator of its faster entry into society 5.0. There are 89 parks in the UK, 36 in Germany and 27 in Spain.

Boredom is related to thinking and all thinking tends to distance us from the real world. Distraction, on the other hand, disturbs reflection. However, the question of knowing why we become bored does not stem from work or free time. Man can be bored at work, or during his leisure time, commonly assimilated to free time. Entertainment might almost seem preferable to the misery of life, because it provides an appearance of happiness for a moment. Expecting to escape boredom through distractions is like trying to escape from reality. Thus, every pleasure brings no more than a passing satisfaction, reduced to nothing by the appearance of a new need. According to Nathalie Joulin, the individual probably makes a distinction between “labor” time and “leisure” time[11].

In other words, one of the characteristics of society 5.0 is the development of a distraction economy. This economy of entertainment is a real industry.

1.6.5. *Which role for identity?*

The current world presents itself as politically correct, smooth, polite[12] (in the moral sense of the word). Thus, for men as for machines, only scars and failures can endow the “body” with its individuality. However, a scar hurts. This is imprinted on ideas as lifetime mark might be.

---

11 Nathalie Joulin, *op. cit.*, p. 18.

12 This refers to the extension of politeness to the whole of society, something which is expected from one person interacting with another one, in terms of behavior, manners and a certain language, everything in conformity with the rules of decorum and the respect for correctness.

Anthropology has brought a strong idea to marketing. Consumption is not only utilitarian, to consume is to produce, meaning that it simultaneously displays an individual identity and a social belonging[13]. This notion of identity is crucial for society 5.0. Both men and machines are distinguished by their identity, one of whose elements is the address.

DEFINITION. An address or an identity is a feature or a group of features whose function is to help locate information (particularly in memory), to identify a recipient or to locate a person or an object.

This notion of identity brings us back to the notions of globalization and regulations.

1.6.5.1. *Globalization versus the local level*

Local products are becoming more and more fashionable. Supermarkets like Cora or Intermarché have always had original products from the village, the canton or the region, but this is also the case for highly centralized stores. Casino offers a palette of more than 700 firms referencing crafted or local products.

The difficulty for defining the local level is that this concept cannot be limited solely to food products, for which the consumer seeks freshness, quality, as well as health benefits. The concept of small circuit is currently valued, and the location of the selling point is an important variable. For example, it is not easy to have local apricots in Roubaix. A local product can convey a positive image or pleasure. Its spread may go beyond the limits of its original region. One of the key elements is geography. While *cassoulet* (a typical French dish) is a local offer in Castelnaudary, it is not in Sarthe. A distinction should be made between local products, which provide one or two stores, and regional products, which can meet the needs of ten or more stores. Local supply can be measured in terms of kilometers. This idea is also present in the notion of "smart region".

A "smart region" is defined by S.M.A.R.T objectives, which can be summed up as specific, measurable, accessible, realistic and temporal ones. The second characterization is the definition of equivalent distance circles, most often in terms of commuting time rather than kilometric distances.

13 Richard Ladwein, *Le Comportement du consommateur et de l'acheteur*, *op. cit.*, p. 29.

These distances are grouped in a unified distance matrix (U-matrix) in order to enable further analysis and the definition of equipment policies.

The influence of the local level reflects the importance of the coexistence with nature, which we had already came across in society 1.0 and society 2.0.

1.6.5.2. *Globalization, massification and individualization*

As a worldwide process, globalization has sought to produce the same product for everyone. Society 5.0 is characteristic for the adaptation of products to individuals via their identity and/or address. Nathalie Joulin has made the hypothesis that global products are becoming more and more personalized. We even witness the development of regional concepts. We multiply contact points with customers. The author has referred to "operational innovation", in which we find new distribution channels and "advertising innovation"[14]. At an identity level, it is more a question of responding to the consumer's boredom rather than bringing forward real product innovations.

One of the features of society 5.0 is that it distances itself from previous industrialization processes, for example, from the industrial revolution that led to single mass production.

1.6.5.3. *Globalization and standards*

Standards are one of the essential elements of this identity, as well as a requirement for this new industry that is being built.

Standards sometimes go through a protocol, for example, a telephone number or an Internet address. Sometimes they are restrictive, as addresses are. The European Article Numbering (EAN) code is a barcode used in commerce and in industry in accordance with EAN International specifications, which associate a code with a product. The same regulation applies to other products: the International Standard Book Number (ISBN) is a unique barcode identifying published books. Some sellers have their own code as well. Thus, the Amazon Standard Identification Number (ASIN)

---

14 Nathalie Joulin, *op. cit.*, p. 29 and following pages.

is the identifier for products sold by Amazon. While for books, the ASIN, the EAN and the ISBN are all the same, this correspondence is not necessarily true for other products.

In the case of identity, we are mainly in the domain of code-information.

1.6.6. *Which role for alienation?*

Boredom always involves the awareness of being imprisoned in a given situation, or simply in the global condition of the world. We could describe this feeling of confinement as "alienation". Alienation cannot be an ideal, since this presupposes knowing what man is alienated from. Alienation makes it necessary to make an explicit reference to another element, the object of alienation. This will probably be the element which has been lost or the one which remained attached to man.

For Bertrand Vergely, alienation characterizes the situation of becoming a stranger to oneself. This situation concerns a human being who has the impression of not being who he really is, despite the fact that he has not lost what constitutes his everyday existence[15].

According to Svendsen, boredom has a dehumanizing effect, which is itself a form of alienation. It removes meaning from man's life, precisely that which constitutes life as such[16]. Man's being might be constituted by the fact of being in the world, that is to say, in virtue of the polarity between subjects and objects, in other words, between man and his surroundings. The same author asserts that, when bored, we experience "the inexistence of reality or the reality of non-existence". Nowadays, this argument is used by medical professionals to fight against the emergence of medical technologies.

Alienation, and particularly technological alienation, is frequently put forward by those who oppose society 5.0, mainly because the former type of society is still profitable for them.

15 Bertrand Vergely, *Petite philosophie pour jours tristes*, p. 76, Milan, Toulouse, 2003.

16 Lars Fr. H. Svendsen, *op. cit.*, p. 184.

1.6.7. *Which role for action?*

The question of action is important in society 5.0 because power is no longer derived from the possession of information, but stems from the ability to act. This is the reason why the transition from a bored consumer to an acting consumer is so important.

In his earliest work, Pascal showed an interest in boredom. In *Thoughts*, he wrote:

> "Nothing is so insufferable to man as to be completely at rest, without passions, without business, without diversion, without study. He then feels his nothingness, his forlornness, his insufficiency, his dependence, his weakness, his emptiness. There will immediately arise from the depth of his heart weariness, gloom, sadness, fretfulness, vexation, despair".

According to Alfred de Vigny, the other remedy against boredom was action, which derived from readiness. In his famous speech at the Chamber of Deputies in 1839, Alphonse de Lamartine reckoned that "France was bored" and that is why it was necessary to have a will to act.

Alain Caillé was interested in the theory of action by Didier Mauss. This theory is structured according to four motivations, organized in two opposite pairs. As regards the first pair, the actor pursues his individual interest, which we can describe as self-centered, seeking personal glory, recognition, and eventually, to please others. These are all free things that cannot exist without the counterpart of a partner, of another one. The teenager's new pair of sneakers with transparent or luminous soles only makes sense provided that other teenagers envy him for these. It naturally follows that he must quickly find another model so as to reinitiate the process.

The second pair of opposites is described by Caillé and Mauss: every action entails an obligatory character related to regulations and social rules. These laws and regulations are opposed to spontaneity, to pleasure and, therefore, to a certain degree of freedom.

In society 5.0, acting becomes a source of happiness, pleasure and freedom.

1.7. The contribution of the first revolution

Jean-Paul Demoule[17] has stated that the invention of agriculture and livestock signified an unparalleled revolution for humanity. The specialized teachings of this period can help us understand the ways in which society 5.0 can be implemented.

Relinquishing the world of hunting and gathering, man was able to take control of a certain number of animals and plants. This represented the invention of agriculture and breeding. In a similar way to the changes which are starting to take place in society 5.0, the agrarian revolution made sedentariness possible. This influenced both mobility and habitat, as it will also happen in the new society. The revolution also caused a demographic boom. As Jean-Paul Demoule pointed out, on average, hunter-gatherers had a baby every three or four years, while women farmers had a baby every year, even if some of the children died at a young age. For this author, the following revolutions had the same consequences. In the context of society 5.0, one of the important issues will be the impact of population growth.

1.8. Humanity 2.0 and society 5.0

In 2005, engineer Raymond Kurzweil published his book *The Singularity Is Near*[18]. In this book, he claimed that, in 40 years, that is to say, in 2045, artificial intelligence would prevail over man. He named this approach "technological singularity".

DEFINITION. In social sciences, singularity is a concept that the American science-fiction writers of the 1950s borrowed from mathematics. It refers to an unprecedented and irreversible global mutation.

According to Ray Kurzweil, networking supercomputers and their software will increasingly become more "intelligent", and end up performing better than humans in all areas. This might signal the advent of humanity 2.0.

---

17 This is the book by Jean-Paul Demoule, *Les Dix Millénaires oubliés qui ont fait l'histoire*, *op. cit.* The author is an archeologist and a distinguished professor at the University of Paris I Panthéon-Sorbonne. He specializes in the Neolithic period and the Iron Age. He was also the President of the INRAP (the French National Institute for Preventive Archeology Research) from its creation in 2002 until 2008.

18 We are referring to the book by Ray Kurzweil, *The Singularity Is Near*, published in French in 2005 as *Humanité 2.0*, M21Éditions, Paris, 2007.

Humanity 2.0 will feature a fusion between man and machine. For this, Ray Kurzweil suggested a step-by-step advancement, which will combine the science and technology of society 5.0: consequences of research on genetics, generalization of nanotechnology and the use of automation and robotics. According to the author, the process will logically start in an unapparent manner: cellular therapies for the treatment of diseases, implantation of electronic chips for analysis and biological follow up, bodily introduction of repairing nanorobots (particularly in the blood stream) and connection of intelligent orthotics and prostheses. From this will follow a "slow but inexorable progression of biological humans towards non-biology". For Kurzweil, even before the change of singularity, "improved humans" will exist. The author even suggested the hypothesis of immortality. With these changes, humans might be able to live well beyond current averages. Our book will analyze the strengths and weaknesses of these technologies.

1.9. The new role of society 5.0: a return to bio?

As society 5.0 has resumed with the codes of the ancient society in its coexistence with nature, it features a "return to bio", that is to say, to organic vegetables, but also to biomedicine, biodegradable plastics, etc. This return to bio is not an ecological doctrine, but it is related to society 5.0 and features significant economic effects.

Bio also refers to biodiversity. This notion is very broad: biodiversity is devoted to the diversity of the living world at all levels: diversity of environments (ecosystems), diversity of species, genetic diversity within the same species, but also the presence of diversity in human and animal actions.

1.10. Growing sectors and lagging sectors

The dynamics of society 5.0 partly stems from economy.

DEFINITION. Cost of Entry (CoE) is a term used for describing the cost of entry of a brand or a firm into a new market: it is related to physical and marketing (advertising, distribution) investment.

Two types of economies are leading the Internet:

– the relocation of traditional industrial economies, particularly with technologies such as 3D printers, technologies that make it possible to reduce the technological entry ticket; relocation is accompanied by huge gains in industrial competitiveness;

– a social and a solidarity economy, related to a “Co” society, which is gradually transforming non-commercial ventures into commercial ones. In this case, we are referring to the marketing entry ticket. We will discuss the concept of “Co” society later on.

1.11. The elements of society 5.0

Adaptability, agility, mobility and reactivity are now key words in the life of society 5.0, which entails the fact that mutations, changes and evolution are an observable constant on a daily basis, which is also reflected by infrastructure, knowledge and skills.

Adaptability, agility and reactivity are very important and require the implementation of industry 4.0, using additive techniques that consume fewer resources for production. The next chapter will be devoted to this new type of industry. Mobility produces effects in transportation and at home, which are increasingly becoming more “mobile” and more intelligent.

Society 5.0 is a fresh “New World” in which the exchange is important. This concept questions the primacy of economic exchanges and the primacy of ideas.

In Western democracies, since it was goods exchange that prevailed, the performance of means of transportation must have favored globalization. However, in society 5.0, it is the primacy of ideas and knowledge that take precedence. Around 1992, China developed its soft power, the primacy of ideas. France is another example, as reflected by its “cultural exception”. Exporting ideas and knowledge have become a source of wealth in a period when goods exports are declining, and service exports are increasing.

The creation of company values does not refer so much to the exploitation and transformation of natural resources, but rather to the use of infrastructure (including transport[19] or telecommunications networks), on the one hand, and knowledge and skills, on the other hand. Infrastructure, knowledge and skills are key resources that need to be managed and which will create the competitive edge that will favor organizational development. As a consequence, some questions arise:

QUESTIONS.

– How do we know which types of infrastructure, knowledge and skills are needed for the organization?

– Which management tools should be implemented in order to manage it?

– Which different types of management and enterprise structure should be implemented?

– What are the challenges the organizations will have to face?

– How will the management of infrastructure, knowledge and skills change collective action?

Our book will seek to answer these questions.

19 This notion refers not only to routes, railways, ports, etc., but also to gas or water networks, etc.

2

From Society 5.0 to Its Associated Policies

Society 5.0 is not only bringing about changes in current policies thanks to the impact of information and the use of data, but also through the technologies that it can embed. In fact, it is modifying the modus operandi of political action.

This type of society might be linked to a political will seeking to attain industry 4.0, which might result from the fusion of digital technologies and the limits of physics and biology.

2.1. The place of politics in organizations

2.1.1. *The three levels: strategic, tactical, operational*

For a better understanding, this section is absolutely necessary insofar as the terms "strategy" and "tactics" have been misused.

The difficulty of strategy stems from the fact that the rules and decisions in terms of research and the evolution of products, markets and technologies differ from one organization to another.

Besides, strategy refers to the activity of the organization itself. The traditional vision of management separates strategy from tactics, and these, from the operational level. The military were the first ones to introduce these concepts.

Figure 2.1. *The three levels*

DEFINITION. Strategy is the definition of objectives and goals to be achieved. It defines the medium- and long-term choices of the firm's battlegrounds. These are generally known as Business Strategic Units (BSU).

DEFINITION. The tactic is the definition of the means to achieve these goals. This is the reason why many authors refer to the organizational level rather than the tactical level. A general in the army only implements tactics and contributes with information in order to define a strategy.

DEFINITION. The operational level refers to carrying out operations, with knowledge of the means and objectives, or goals, to be attained.

2.1.2. *Politics and ethics*

Politics and ethics are situated upstream of the strategy-tactics-operations trilogy. At the top of this vision, there is a political and cultural framework. For this reason, we often consider the management approach as divided into four main stages.

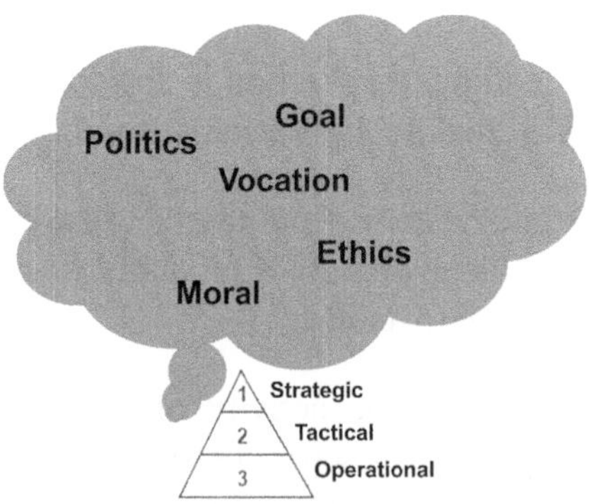

Figure 2.2. *Political and cultural cloud*

In a certain way, this cloud of politics and culture is the identity card of the organization. Despite the fact that this framework is vague, it still exists and it delimits the possible actions of the firm. The firm's strategy must therefore fall within this field.

2.1.3. *The relationship between the strategic, tactical and operational levels, and the organization's functions and tasks*

Stages	Functions	Tasks
Political and cultural	-	Observation Analysis
Strategic	Direction	Organization Coordination Control "Major choices" Planning
Tactical	Supply Financial, human investment	-
Operational	Production Commercial	-

Table 2.1. *The different organizational levels*

The marketing and sales function is a concept, a vision whose reasoning is represented as opposed to the previous system. As a result, it goes through all four stages.

2.2. The implementation of national policies

Regarding the implementation of national policies in this field, different theses confront each other.

The supporters of these new technologies consider that the advent of industry 4.0, and therefore, of society 5.0 is unavoidable. Nicolas Bouzou and Luc Ferry are examples of this in France. Some others, such as Ray Kurzweil, whom we have already quoted, go even further. For these authors, it suffices to describe its dawn.

For some other authors, such as Erik Brynjolfsson and Andrex McAfee, this revolution generates inequalities and value upheavals, especially regarding work. These changes increase inequalities. Among these, some believe that governments will adapt to this situation. According to John Micklethwait and Adrian Wooldridege, this adaptation will have to face the pitfall of populism, a crisis of representativeness and the search for the efficiency and performance of public authority[1]. Their book is called *La Cinquième Révolution: la course mondiale pour réinventer l'État* (The Fifth Revolution: the world race to reinvent the State). They are confident because advanced nations have always invented new government forms in order to adapt themselves to technological transformations.

Different points seem important:

– governments will have to accept the citizens' aspirations and blow up the walls that hinder them from achieving these;

– the current forms of State are not performing well enough due to financial deficits, obsolete social regimes and the distance of elites from the rest of the population. It will therefore be necessary to restore an effective State that respects individual rights and is based on the values of parliamentarism. Besides, the budgets of international firms are sometimes higher than those of States, which is a source of difficulty;

– the obesity of the State is not the problem. We must separate the question of the State from the public sphere, and define real public service missions, which may not be guaranteed by the State, but by associations, citizens, consortia and even private firms. It is a question of deconstructing the walls of public management and the civil servant status, without seeking to remove these rules, but by adapting them to society 5.0 (the different types of walls will be studied later).

States will have to become role models themselves, since it is unlikely that international organizations will do so, considering that their role will be restricted to pointing out the good examples.

1 This refers to the book *The Fourth Revolution: The Global Race to Reinvent the State*.

2.3. The notion of walls

The notion of walls is a fundamental part in the development of society 5.0.

DEFINITION. Walls are groups of people or groups of individuals who block the development of society 5.0 in virtue of their functions or their power.

2.3.1. *Different types of walls*

It is imperative to break these walls in order to enhance the development of new societal forms. The theorists of society 5.0 mentioned the existence of six types of walls:

– the wall of the government and its agencies;

– the wall of the legal system;

– the wall of technological innovation;

– the wall of human resources;

– the wall of access costs;

– the wall of user acceptance.

Regarding the first wall, a solution could be to ask politicians to create plans and to turn them into laws, which would certainly have an impact on the second wall, that of the legal system.

The legal system wall can be overcome by inviting the legislative system to exclusively issue laws that promote innovation. It might also be necessary to prevent the enactment of laws protecting an industrial sector or forbidding innovation. The precautionary principle, which is present in the legal arsenal of many countries, is a clear example of the laws to be excluded. Among the walls of the legal system, there is a very important one: that of the particular and professional limit.

The technological wall can be overcome thanks to education. The idea is to train a group of individuals of a specific age range with the necessary elementary technological concepts. This could serve as an educational reform, which helps to integrate this type of knowledge, and sets up an evaluation mode, which could assess progress in learning.

The human resources wall can be lifted by eliciting commitment from all citizens to create this new society. One way could be tax incentives, while others could include labels, contests or awards.

The wall of access costs is one of the most important ones and should not be confused with that of social acceptance.

The wall of social acceptance is the most complex one. The goal is to ensure that all systems are accepted. For this, it is necessary to implement design and ergonomics methods, as well as citizen participation in the co-design of products and services.

2.3.2. *The "NIMBY" wall*

Other walls exist, such as the NIMBY, or "Not In My BackYard" problem.

DEFINITION. NIMBY is the acronym for "Not In My BackYard", a term generally used for describing either the opposition of a group of people to a general interest local project, which they consider to be potentially harmful, or to the neighboring residents of the implemented project.

Everyone wants to enjoy high-speed train lines, but they do not want them to be located near home. The list of actions of this type could fill up an entire book, but we have to bear in mind an important element. The most common effect of NIMBY shares is that they delay the arrival of projects, whereas society 5.0 is characteristic for its need for agility. It takes an average of 6 years to set up wind turbines somewhere in France, despite the ecological will of politicians.

In the same way, all the current mayors of big cities desire to implement the electric car, but with the idea that electricity production is done elsewhere. Paris is an excellent example. For a French car fleet of 39,140,000 vehicles, around 4 million cars run for 33 minutes per day in the Parisian region, which corresponds to 2 million hours of traffic, an average power of 113 horses, that is to say, 84 kWh. Taking losses into account, it is necessary to store about twice as much energy, for instance, about 150 kWh, which would correspond to the consumption of more than 300 megawatts, namely one third of a nuclear power station!

2.3.3. *The wall between private individuals and professionals*

This wall was addressed in the section regarding legal walls. It is very important because, in many countries, it reflects changes in the rule of law, particularly in the transition from consumers' right to commercial law, even if, in our evolved nations, these two levels of rights have different and sometimes opposing rules.

Let us take the example of apartment rentals. In France, there is a threshold in France called "the 120 days". If the main residence is rented for less than 120 days, there is no procedure to carry out, according to the ALUR law of March 24, 2014, regulating the access to housing and renovated town planning. However, there are exceptions already, such as professional obligation, *force majeure* or health reasons! If the property's function changes after 120 days, a declaration must be made at the city council. The process has become mandatory in Paris, Bordeaux and Nice, and inspectors are analyzing the advertisements. A change towards the professional world is subject to authorization in Paris, Lyon, Marseille, Nice and Strasbourg. In Paris, this rule is accompanied by a compensation obligation, when property is purchased in order to turn it into dwelling premises. In principle, this tenancy is prohibited to the occupants of social housing or assisted private housing, despite the fact that everyone knows that housing seaside towns are rented throughout the summer, and paying an annual rent to their beneficiary.

The city of Barcelona wants to fight against a double fraud to Spanish law. The scam refers to renting a flat for a long-term period, by means of the traditional commercial transaction, and then subletting it to tourists for a few days, via a platform like Airbnb, Abritel Home Away, etc. The Barcelona city council has decided to curb the effects of mass tourism in its Catalan capital, seeking to fight against this double fraud. More than 3,000 illegal tourist apartments were detected in early 2017 before the tourist season, including over 300 sublets. In the Belgian Flanders, although a regional law exists, De Tijd newspaper estimated that 85% of the advertisements did not respect the regional laws. The law requires web hosts to record their property at the city council and to include their identifier on the advertisement[2].

2.4. New political attitudes

New political attitudes can be observed: let us point out vetocracy and ultrademocracy. These new attitudes work as walls in the sense that they slow action down.

2.4.1. *Vetocracy*

Francis Fukuyama has published numerous books on the State of nations and the decline of institutions, particularly in the USA. However, his books also deal with the economic development associated with this. We can thus assert that the author offers a tool for reflection on the possibility of a new industrial society.

For Fukuyama, political decay is the result of the "repatrimonialization" of government institutions. According to him, this phenomenon is a profound cause of political decline.

The phenomenon first occurred in the USA during the second half of the 20th Century. From a historical point of view, it might have been present at other periods: at the end of the Han dynasty in China and under the *Ancien Regime* in France.

Repatrimonialization can be identified thanks to certain indicators:

– sponsorship within the public service;

– political clientelism (sectoral subsidies, targeted tax credits, tariff barriers);

– institutional patronage (distribution of job offers to friends and relatives of those in power);

---

2 For further information: http://www.lemonde.fr/les-decodeurs/article/2017/08/04/louer-sur-airbnb-un-business-lucratif_5168810_4355770.html#7j14BccHImiisF4T.99.

– the absence of more coherent citizen representation, which leads to a unique solution: permanent citizen protest, either through associations or in the street.

Permanent citizen protest is the most prominent indicator, according to Marcel Gauchet[3], historian and philosopher.

The phenomenon of "repatrimonialization" has several consequences:

– the extension of bureaucracy through the creation of superfluous agencies and staff positions;

– the transformation of public services into inefficient activities, sometimes paradoxically carried out by private groups;

– the capture of political life, and especially of political decisions by interest groups which are both organized and funded, and often not very transparent;

– the interference of justice into the legislative function, or even in administrative tasks;

– the entanglement of legislative texts produced by the legislator, sometimes with contradictory injunctions making government functions futile.

DEFINITION. Vetocracy appears when the political system grants an effective right of veto to a large number of actors.

These actors are generally involved in a conflict of interest, which blocks any public action or social innovation forms. Vetocracy is in contradiction with the existence of society 5.0 on many points. More and more voices are rising to defend an idea related to vetocracy, whereas it opposes the very same concept of "Co" society: according to this movement, citizens should be able to irrevocably oppose a decision made by their elected political representative. Some have even suggested adding a popular referendum to the legislative process, which could challenge the decisions of a democratically elected government. For Marcel Gauchet, representative democracy, participatory democracy and parity democracy are dangerous ideas which can lead to constantly blocked, paralyzed power, incapable of

3 For more information : https://www.publicsenat.fr/article/politique/de-la-democratie-a-la-vetocratie-selon-marcel-gauchet-53301.

making decisions. In fact, "Co" society (related to economy) should not be confused with political decision, which is the action of politicians related to well-trained administrative civil servants.

For Marcel Gauchet, "behind the demands for participation, the ability of elected representatives to validly represent popular will is challenged". He adds: "Every participatory experience leads to the same conclusion: that this participation is not so prominent".

The existence of bad governments does not stem from the absence of participatory democracy, but from a lack of autonomy. This results in a public service governed by strict and inflexible rules, whereas it should:

– interact with the political injunctions of governments;

– contribute to decision-making, and to the construction of strategies;

– implement effective means, which means having good tactics;

– listen to the economic actors set in this "Co" society, so as to be as operational as possible.

At this point, it should be noted that the rules built by the civil servants themselves not only proliferate, but are of such a nature that their full application becomes impossible. This results in arbitrary powers given to the administration. In this way, every actor can decide which rules apply to whom. On the other hand, for the administration to produce effective actions, it must:

– be autonomous from political powers;

– be judged by the consequences of its own decisions.

Evidently, the latter produce consequences in the economic world.

In order to develop administrative efficiency, it is necessary to have a higher education system capable of training administrators properly. The *École Normale Supérieure* and the *École Polytechnique* were founded by the Revolutionary Government of France in 1794, with the aim of training competent civil servants. This trend was followed after the Second World War by the *École nationale d'administration*. The Stein-Hardenberg reforms in Germany (1807) or Northcote-Trevelyan in England (1854) and the Pendleton Act (1883) in the USA were similar initiatives and highlighted the skills of a proper administration, denying patronage.

This administrative efficiency makes it possible to put private actors in good hands, and is therefore a factor to be taken into consideration in the search for industrial competitiveness.

2.4.2. *Ultrademocracy*

"Ultrademocracy" is a term proposed by Dominique Schnapper[4] in her book *L'Esprit démocratique des lois*[5]. The question that this specialist in democratic dynamics intends to analyze is the extent of individual claims or of claims made by groups of individuals. In other words, where do the extremes of the nation and the individual meet? She is interested in the contradictions and tensions that arise within the democratic system, including the risk of "corruption". She characterizes the current State as suffering from "democratic discomfort", which persists despite more transparency, populations have more liberties, morals are more tolerant and societies are economically richer than ever before. However, transparency is one of the foundations of factory 4.0, which will be described later. One of the first solutions the author is considering is to anticipate problems, because all observers know that laws are always lagging behind reality. To illustrate this, she quotes many examples.

The difficulty of current democracies seems to be the willingness of individuals to always satisfy their material interests by concealing the "common good". Could individualism be a cause of democratic disintegration? The author regrets the disappearance of a republican conviction in political ideas and major projects. The problem of democratization could stem from this civilization of immediate and individual convenience. According to this sociological vision, the individual has replaced the citizen. The consumer society is replacing the political community.

The second point developed by Dominique Schnapper is the "temptation of the unlimited é". This temptation should be associated with free economy,

---

4 Dominique Schnapper, was a member of the Constitutional Council of France (2001-2010), and appointed by the French government since December 2017 as President of the Secular Elder Council for National Education. She is the daughter of Raymond Aron, a well-known French philosopher, sociologist, political analyst, historian and journalist.

5 We are referring to the book by Dominique Schnapper, *L'Esprit démocratique des lois*, Gallimard, Paris, 2014.

often promoted in society 5.0. While man is subject to biological and physical limitations, technological and scientific progress gradually seem to introduce the idea that one day it will be possible to eliminate boundaries. The author puts herself in a critical position facing the thesis of humanity 2.0 and singularity. By extension, citizens might be convinced that the democratic society has conquered Marxism and cannot be extinguished. The consequence is immediate: "citizens" do not take care of democracy and mainly focus on the satisfaction of their personal interests.

Citizens seem to reject rules, laws and social regulations. They challenge the role of those who are responsible for enforcing them. For this reason, Dominique Schnapper proposes ultrademocracy or "radical" democracy.

DEFINITION. Ultrademocracy can be described as an evolution of the democratic society that erases the concept of authority related to age, experience, competence and institutional weight.

This form of democracy features an adherence to the concept of society 5.0 and to the idea of "Co" society. Questioning authority seems to be spreading in the democratic society and, according to the author, this should be put to an end. Democracy is fragile because its functioning requires learning. Institutions and organizations should be simultaneously strong and flexible. In accordance with society 5.0, the most proper term would be "agility".

2.5. The role of governments

The role of the State in society 5.0 is multidimensional. First, it seeks to protect national industry, and sometimes this produces unexpected twists that lead to misinterpretations.

2.5.1. *The protection of national industry*

In this world of competition for finding new customers, States most often seek to protect their national players.

China is often mentioned by the Americans to illustrate this topic. This is particularly the case of the Alibaba online commerce site, which carries more transactions than eBay and Amazon websites, although Chinese assets are in

Amazon. In this country, Wechat messaging reaches one billion subscribers. Uber was expelled from the Chinese market by Didi.

Europe has tried to encourage the birth of a competitor to Google, but without success.

Sometimes protection works in the opposite direction, as States try to protect old industries against new industries.

2.5.2. *The limitations required by governments*

Some States have advocated for a limitation of certain functions to specific digital players and firms using these technologies. These actors communicate promptly in relation to these topics. An article in *Le Monde* from August 8, 2017, was entitled "GAFA facing the Great Wall of the Chinese State". This title resumed the notion of wall.

For example, the great American web services are blocked in China, because Facebook, Google, Instagram and Twitter do not conform to Chinese laws. Sometimes, it is just a few pages of the website which are not visible in certain countries. However, our Western countries are not exempt from these practices.

Apple has removed virtual private network solutions (VPNs) from its applications because these solutions made it possible to circumvent controls. Some people are surprised that Apple is subduing to the Chinese State in virtue of economic interests, and that the firm refuses to unlock iPhones belonging to terrorists in the USA, arguing the respect for privacy principle.

The mailboxes of local dissidents in some countries are being attacked. Google had a negative experience in China. Some important actors, including Google, have pulled out of the market. Facebook, in contrast, has modified its website to make it compatible with Chinese activities.

Contrary to spread ideas, this type of attitude is not only due to China. In most European countries, Facebook has removed its search engine to comply with computer and freedom commissions. This feature exists in other national versions and it is enough to declare that one is American to access it!

These limitations only make sense if States wish to limit shocks associated with these breaks by "softening" inevitable evolution.

2.5.3. *The question of public orders*

The allocation of public orders reduces competition. In this case, the State plays the role of market arbiter. Corporate success or failure owes much more to government interventions and their ability to make the best use of these than to the competitive strategies that firms are likely to implement. On the other hand, export competition is strongly influenced by the laws of the market.

2.5.4. *New cultural policies*

Cultural policies are receiving the impact of technologies of society 5.0, and are consequently involved in the changes related to this new type of society. However, cultural policies are also at the heart of the entertainment industry which we have described. Let us give some examples.

In South Africa, the Yoza project is very interesting because of the lessons it entails. Initially called "m4Lit" (mobile phones for literacy), the project aimed at making people who do not frequently read, to read more. Thc solution requires the use of a mobile phone. The approach involves lessons, but in the context of a "Co" society. The approach implemented in the mobile application makes it possible to engage the audience, to quickly renew the contents, and to bear in mind the fact that the mobile phone is always connected. Users know what they want and how express it, users want to make comments and vote, and existing networks are put to profit, by becoming vectors for communicating the solution. The rating system makes it possible to know which "texts" please the most. Despite the fact that for a given user performance erodes over time, there are some loyal fans who love to express themselves, as well as many silent users. The question then is that of the "relaunch" of usage. This example is also revealing of the importance of the knowledge society. The economy of knowledge has given rise to very useful skills for society 5.0.

Tuareg communities exchange music using mobile phones. Trading conditions are person-to-person, usually via Bluetooth, but other exchanges take place via networking. The link between technology and use is perfectly obvious. This is also another example of the exchange society or the economy of exchange.

3

Industry 4.0 at the Core of Society 5.0

The factory of the future will be a factory first of all! It will not be virtual, as some so-called experts have imagined, but will take many forms. This book aims to describe this factory, as well as to explain it. The factory is a firm whose outline needs to be delineated. In the following sections, we will describe what a factory is like, or rather, what the "factory" species is like, and its different types.

Let us ponder the definition of these words. The term "species" comes from the Latin word *species*, which designates a "type" or "appearance". This appearance is currently changing and the big factory chimney spewing black smoke is slowly disappearing from our landscapes. On the contrary, it is easy to locate telecommunications rooms with huge window panes, which match the equipment installed inside these establishments accordingly. The notion of "species" is a theoretical one, and is a term that people employ to describe things. Nowadays, datacenters have no windows and rather look like bunkers.

DEFINITION. A species is a group of individuals, which are similar in shape and composition.

According to the definition coined by the life sciences, every representative of a species is alive and reproduces itself! What could be better for the factory of the future? "Genres" represent the different variants of the species when standards are imposed.

Let us follow Plato's logic for a moment. According to this Greek author, insofar as we see horses and cows, but never hybrids of these two races, in some place there must be an ideal form which constrains an animal to be one or the other. We are currently describing these ideal factories, which are plainly "cows" or "horses". We will display a portrait of forms, we will anticipate the different types of factories of the future, and we will see especially to what extent standards become necessary.

3.1. Business in society 5.0

This section is devoted to business, which needs to evolve towards industry 4.0. For this, we will first analyze the decline of industry, which encouraged the leap towards industry 4.0.

3.1.1. *The recent history of the decline of industry*

The recent industrial decline in Europe and in the USA was acknowledged behind schedule by public policies. For this reason, it has encouraged the birth of plans for industry 4.0 or for the industry of the future in several European countries. The French industrial decline is not recent. The 1990s witnessed a sharp drop in manufacturing output: –12.5% between the first quarter of 1990 and the fourth quarter of 1993. The same happened in other European countries. The decline had begun earlier in Britain, but was less noticeable in Germany.

Between 1990 and 1993, the economies of the advanced countries went into recession. This recession was amplified in many countries due to the unfortunate tightening of monetary policy, which resulted in negative elements. Interest rates were blazing. As a consequence, households and businesses had to reduce their debt as a matter of urgency, and the construction sector reduced its production. This had an important impact on industry and accentuated its downfall.

Despite the termination of this period, the serial devaluations in the UK, Italy and Spain provoked a loss of competitiveness in some other countries, such as France, and jeopardized large parts of their industries, in particular regarding consumer goods: textile-clothing, leather, footwear and household appliances. On the contrary, those industries were reborn in countries where textile firms such as Inditex (Zara) flooded Europe and where the Mondragon cooperative, producer of appliances, bought the brands without restarting the factories. The fiscal tightening of Juppé's government, including a two-point increase in the VAT, slowed down household consumption.

A decrease in the prices of raw materials and oil, and the growth of global demand led to an industrial upturn in early 1997. The "new economy", stemming from the digital, also led to an increase in manufacturing output, supplying telecommunications and computer equipment. The euro was worth less than a dollar and Germany had to pay a high price for its much desired reunification. When it was launched on December 31, 2001, at midnight, the euro was worth two German marks.

3.1.2. *The impact of political choices*

The political choices of the period when the euro began aimed at supporting household demands: consumption and real estate. As a result, politicians lost interest in the industry and its factories.

Different factors can be added to this:

– fabless companies, or an industry without factories, were a concept highly in vogue – this text will explain later why this was a mistake;

– the rise of emerging countries and quality products produced in China. Nowadays, China is seeking to facilitate the export of its products through railway silk routes, which are faster than maritime routes;

– the outsourcing of service activities, massively outsourced to subcontractors. Accounting is a good example. Large groups automate accounting by using bank flows and relocate manual accounting activities, particularly in eastern European countries such as Hungary and Poland.

The bursting of the bubble of the new economy went hand in hand with the entry of China into the World Trade Organization (WTO) at the end of 2001. An acceleration of globalization was one of the first consequences. Some European multinationals relocated their production and others preferred to buy items in emerging countries instead of manufacturing them. Industrial groups also lost control as a result of globalization. Entire sectors fell under the effects of a takeover by a foreign group, even (and particularly) if they were flourishing. In France, this was the case of the aluminum sector, with the disappearance of Pechiney in 2003.

The determinants of competitiveness were questioned around 2007 and 2008 by the early stages of society 5.0, which affected all countries, including Germany, China or Japan. It was during this period that the first reflections on industrial competitiveness were born.

The slump in domestic demand as a result of austerity cures in Europe did not make the resumption of industrial production any easier, except in a few sectors. Ten years later, the tax measures executed did not favor the desired restart. This was accompanied by the introduction of a new policy, called "supply policy", and from 2015 onwards, recovery plans were dedicated to the industry.

3.1.3. *Pierre Musso's perspective*

In his book *La Religion industrielle*[1], Pierre Musso is interested in the evolution of the world of work and business. In order to support his reasoning, he builds a genealogy of the industrial and business religions, by means of three ideological and technical bifurcations incarnated in three symbolic places.

For this author, the monastery premises (11th-13th Centuries) offered solutions, which have remained to this day (wine, beer and abbey cheese). Manufacturing (17th-18th Centuries) replaced this structure to fabricate the first tools. The latter will be replaced by the factory (19th Century), which will try to produce goods in a more economical way, using machining technology, the assembly line and Fordism. Then came the firm (20th-21st Centuries), dominated by branding and marketing, with production becoming partially or totally outsourced. It is necessary to mention that this breakdown differs from the one traditionally introduced in the development of society 5.0.

The digital revolution, that of information technology 2.0 which will be discussed later, further damaged the position symbolized by the firm, or even the office. The metaphor of the desktop is used for designating the computer's home screen, and the screen is supposed to mirror the desktop; it is a kind of Digital Twin. The question of the introduction of mobility 3.0,

1 We are referring to the book by Pierre Musso, *La Religion industrielle. Monastère, manufacture, usine. Une généalogie de l'entreprise*, Fayard, Paris, 2017.

which will be studied later, also questions the firm as a place. Industry 4.0 emerged with the "Managerial Revolution", cybernetics and digitization. It is a combination of connective and digital tools: a phone, a computer, and a 3D printer, all virtual storage places for a just-in-time business. Later on, we will discuss the importance of localization.

Society 5.0 might belong to a new industrial religion, an industry whose symbol might be the absence of a symbolic place, but also that of production carried out closer to the consumer.

3.2. The firm: a general theory

At this stage, it is important to understand the different visions of the firm. We must also agree not to criticize the firm for what it is.

DEFINITION. A firm is primarily a system whose function is economic. The first aim is to generate profit, which is the difference between revenue and fixed and variable costs:

$$\text{Profit} = (\text{Price} \times \text{Quantity}) - \text{Fixed Costs} - (\text{Variable Costs} \times \text{Quantity})$$

This vision is called "economic purpose of the firm".

Fixed costs are essentially the investment that needs to be financed. Profitability is the sustainable nature of profit. Today's funding is generating the profit of tomorrow. It is important to generate a margin, that is to say, profit from which to deduce financing means.

3.2.1. *The management of a firm*

For the most part, this is all about producing. The definition of management is linked to that of the firm and its purpose.

DEFINITION. Management is an American term reflecting the "handling" of people, their behavior and their direction.

From the theoretical point of view, this theme was developed in 1920 by Fayol and Moreno.

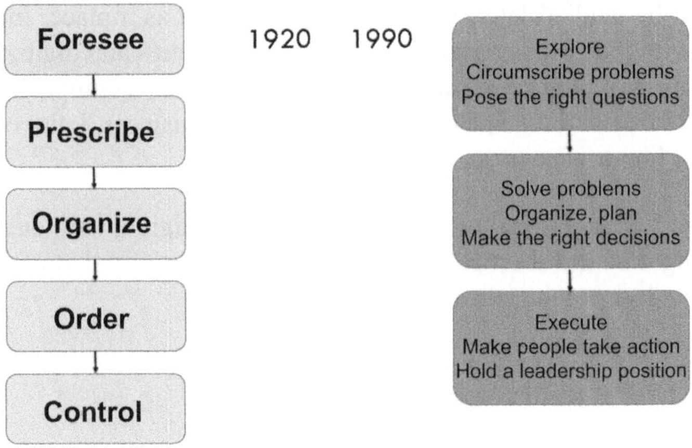

Figure 3.1. *Fayol's budget planning in 1920 and its modification in 1990. For a color version of the figure, see www.iste.co.uk/salgues/society.zip*

Management is based on four elements.

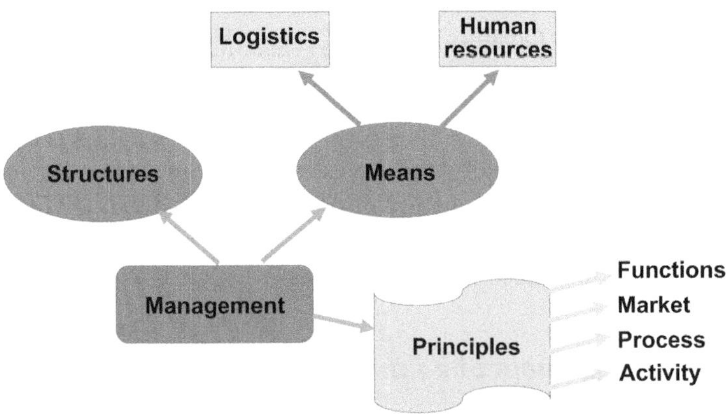

Figure 3.2. *Management components. For a color version of the figure, see www.iste.co.uk/salgues/society.zip*

The economic purpose of the firm and of management does not make sense unless a market exists, even in society 5.0. The development of products that appear to be free for the consumer does not make it possible to avoid the existence of a market. It is therefore necessary to define and evaluate the different historical stages of the market.

3.2.2. *The definition of a market*

DEFINITION. The market is a meeting point between supply and demand that is undergoing changes in management, and which itself results from changes in the environment. In Western Europe, the majority of authors have distinguished different phases in this evolution.

Business type	Period	Marketing features
Taylorist firm	Previous to 1945	Demand > supply Production increase
(Fourastié: The Glorious Thirty) Market research Segmentation	1945–1975	The Glorious Thirty Marketing
Customer's will Cost and economic conditions	1975–1990	The Fifteen Dark Finance and management control
Supply > Demand	1990–2000	The authentic years Personal management

Table 3.1. *Different marketing stages according to the market's condition*

These different stages may vary from one product or service to another.

3.2.3. *The concept of productive activity*

This concept is related to firms. When analyzing management, it is a question of exploring how productive activities can be organized in harmony with the social demand for goods and services via certain organizational mechanisms (such as the market) and how variations and imperfections may affect efficiency. Firms seek to be efficient.

Each economy must decide what end products it wishes to produce and how many of them. There are several alternatives to solve this problem:

– first of all, decisions can be made according to tradition (case of the Middle Ages);

– second, decisions can be taken in central planning bodies (Eastern countries before opening up);

– finally, decisions can result from the market: in the latter case, the key to success lies in the definition of strategic activity areas made by the firms;

– in society 5.0, decisions come from data batches (*datum*).

The approach of the industrial organization takes place in the third solution, the first two types of decisions being excluded. Marketing should therefore be understood within the context of Western industrial economies. It has increasingly abandoned the study of the labor market and that of the banking sector as a mode of financing, and is currently focusing on manufactured goods, mineral products, retail, transportation and energy production. The variables that we now pay attention to on the markets are prices and produced quantities.

The industrial organization has defined "standard" structures for organizations.

3.2.4. *The fundamental structures of the firm*

Different business theorists have defined fundamental structures. These can be challenged by the arrival of society 5.0 only if information flows or the exchange of data are inefficient. On the other hand, it is likely that organizations will have to change structures because of these flows and exchanges.

The fundamental structures result from the analysis of many firms and the attempts to classify them.

3.2.4.1. *The hierarchical structure*

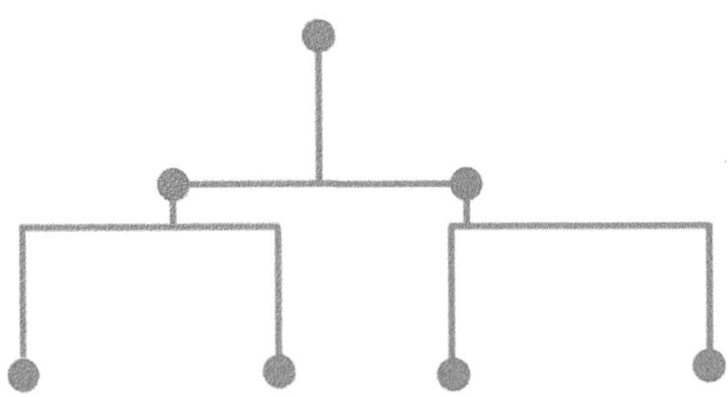

Figure 3.3. *Hierarchical structure*

The hierarchical structure is the historical structure of the firm. It is characterized by the absence of economic logic in its organization, and rather reflects a historical or family logic.

– Advantages: little external vision on its functioning, little control over the firm's strategy.

– Disadvantage: authority structure is often broken.

In society 5.0, this structure can be effective if feedback development is based on synthesis reports, for which the concept of dashboard is really effective.

3.2.4.2. *The functional structure*

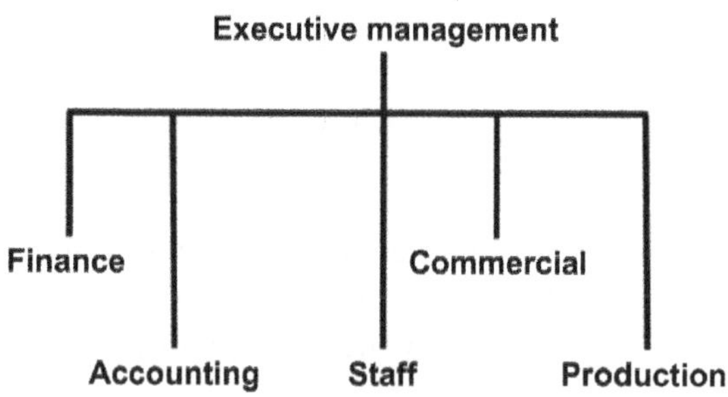

Figure 3.4. *Functional structure*

The functional structure is an evolution of the hierarchical structure. The function serves as a foundation for the structure.

– Advantage: defined tasks.

– Disadvantages: management problems, production/commercial dilemma, "conflict".

The functional structure is the least suitable for society 5.0.

3.2.4.3. *The divisional structure*

With divisional structures, the firm is divided into consistent units. These units are often referred to as "strategic units" (SU) or "strategic business units" (SBU). Two modes are possible, and they correspond to an organization by product or by market; in fact, there may be more complex structures (divisional ones).

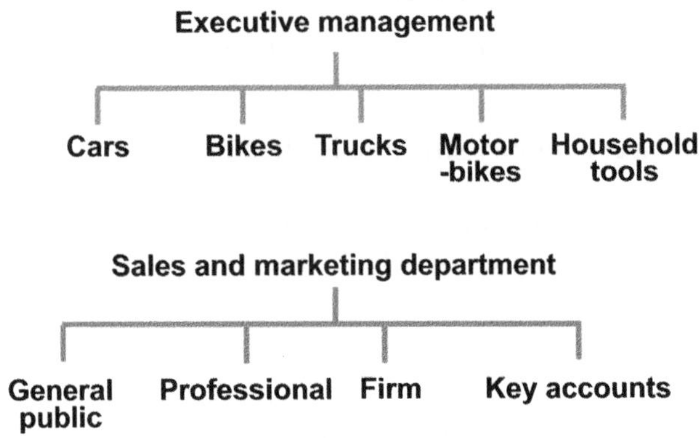

Figure 3.5. *Divisional structure. For a color version of the figure, see www.iste.co.uk/salgues/society.zip*

The divisional structure has the same advantages and disadvantages as the functional structure. Regarding society 5.0, they can be efficient if the units defined in this way are clearly independent of each other.

3.2.5. *The question of the appearance of improved structures*

Improved structures appeared in the early 1970s in order to try to solve the beginnings of the crisis. They featured a mixture of two fundamental structures, so as to mitigate defects and improve performance.

3.2.5.1. *The matrix structure*

The model of the matrix structure was promoted by Anglo-Saxon consulting firms in the 1970s.

– Advantage: creation of relays when the authority relationship is weak.

– Disadvantage: blocking power takeover.

For example, in the 1970s, the great engineering schools reorganized themselves with directorates by year and disciplinary groups of professors.

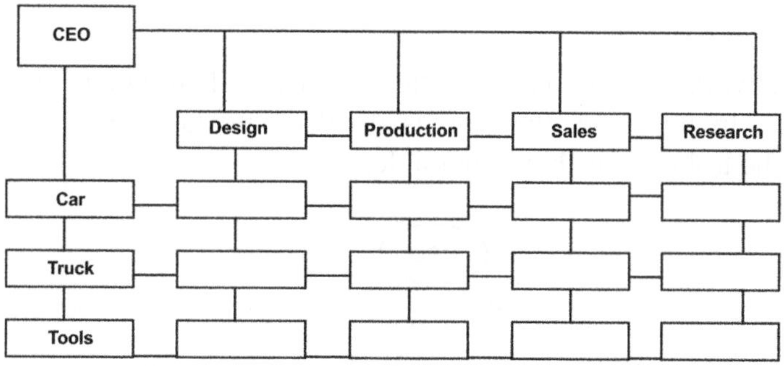

Figure 3.6. *Matrix structure*

One of the expected advantages is that management does not become involved in the content of the rows/columns and that balances in resources are achieved naturally. In real practice, this has not usually been verified. The matrix structure is often presented as the structure of weak or poorly charismatic leaders.

In society 5.0, matrix structures were introduced as the ideal ones, for some time. It was enough to be able to manage two superimposed flows, those allowing the matrix structure to work and those which make it possible to optimize the managerial functioning of the organization.

3.2.5.2. *The "Staff and Line" structure*

The "Staff and Line" structure is a combination of the functional structure and the hierarchical structure.

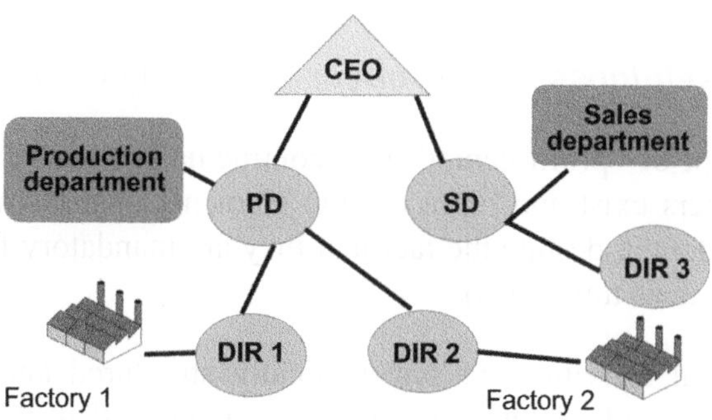

Figure 3.7. *"Staff and Line" structure. For a color version of the figure, see www.iste.co.uk/salgues/society.zip*

3.2.5.3. *The network structure*

The network structure is also called "Structure Task Force" or "Network Structure". In this case, the firm is divided into tasks, so that the links between the tasks are as thin as possible.

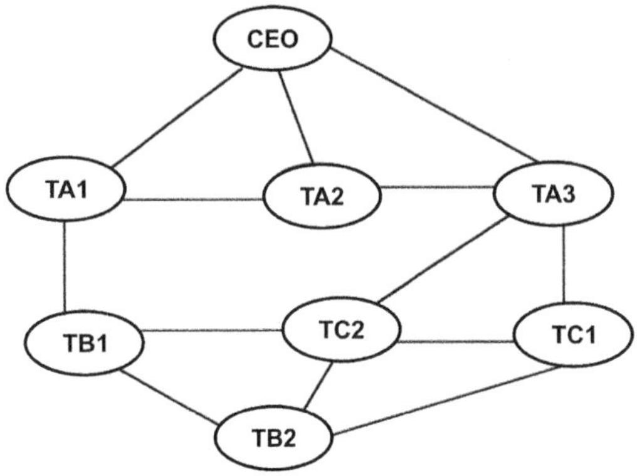

Figure 3.8. *Network structure*

The network structure has been presented as the ideal structure for society 5.0. In reality, the implementation of such a structure comes up against the definition of the organization's network elements or network nodes, as well as the setting up of relationships. Later, certain actors proposed observation construction methods, also called "natural" methods. These include approaches featuring sociocracy or holacracy, which will be discussed later in this book. These are considered holistic views of management.

3.2.6. *The usefulness of the concept of profit center*

The concept of "profit center" is a contractualization between services. Resource centers exist and represent the elements of the structure that are constantly in deficit, despite the fact that they are mandatory for the survival of the firm (such as training is).

All technical activities are systematically structured and managed: by pooling human and material resources, verifying that they generate a turnover and a margin.

The head of the profit center brings his expertise to the sales force in the development of commercial offers. He sets up the necessary structures for carrying out activities and ensures the quality of these. He manages the operating account, not the balance sheet: in that case, we speak of internal marketing.

3.2.7. *The difference between functions and structures*

We should not confuse function with structure. The function exists independently of the firm and its size. The structure depends on the firm and changes as time goes by, according to the size of the firm. In this case, there is a search for competitiveness, that is to say, a resort to flexibility and adaptability.

In these three visions (environment, function and structure), the importance is coherence. It grants the survival of the firm and the pursuit of its goals. This situation is all the more vital in society 5.0.

3.2.8. *The relationship between environment, strategy and structure*

The work carried out by Chandler, a professor at Harvard Business School, has led, among other things, to the following diagram:

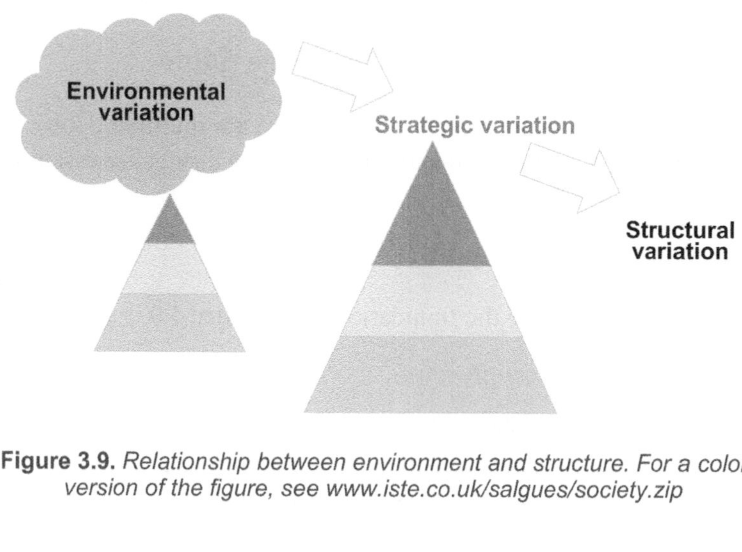

Figure 3.9. *Relationship between environment and structure. For a color version of the figure, see www.iste.co.uk/salgues/society.zip*

This relationship is known for the adage that results from it: structure depends on strategy. This concept was criticized by many authors, particularly by those who intended to rationalize budgetary choices. These same authors insisted on the contingency of the structural reality of a firm. The underlying idea is the challenging nature of the environment, which actually introduces excellence or performance criteria.

Chandler's relation should therefore be represented as in Figure 3.10.

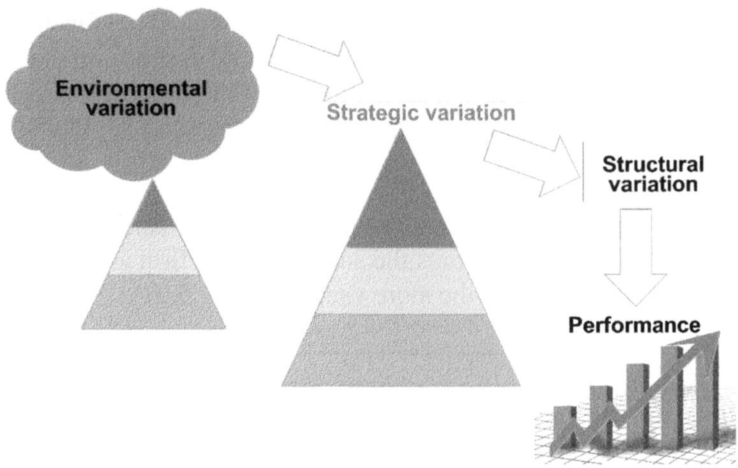

Figure 3.10. *From environment to performance. For a color version of the figure, see www.iste.co.uk/salgues/society.zip*

3.3. The determinants of the factory of the future

The determinants of the factory of the future are multiple. However, two of them, in this case digitization and direct manufacturing, are essential.

3.3.1. *The main determinants*

Four basic elements are the foundation for industry 4.0.

3.3.1.1. *Decision decentralization*

The ability of cyber-physical systems to make decisions by their own means makes it possible for decisions to be decentralized. In this way, tools related to men can perform their tasks as autonomously as possible. A kind of decision decentralization is taking place. Cases of recovery by men are more an exception than a rule. Most often it is a question of interferences, actions, contradictory objectives and constraints to the execution of delegated tasks which come from a different level of the firm. For example, these could stem from a modification in the product manufacturing launch schedule, related to a customer prioritization request by the sales department.

3.3.1.2. *Interoperability and automation*

The interoperability of machines, devices, sensors and people has become global. The industry 4.0 factory provides the possibility of connecting physical and human elements. The latter also make it possible to communicate with each other via the Internet of Things (IoT) or the Internet of people. This results in greater automation of production processes.

3.3.1.3. *Information transparency and virtualization*

Current information systems have the ability to create a virtual copy of the physical world. They enrich digital factory models with sensor data. Afterwards, it is necessary to be able to aggregate these sensor raw data with the orders coming from humans, on the one hand and with higher value information related to the context in which man and the machines are located. However, current information systems, whose theoretical bases date from the 1980s, are based on "context free" languages and only awkwardly admit the introduction of the conceptual.

The transparency of information and its sharing is a necessity in society 5.0 because this society is based on the exchange of information. As we will see later, and this is barely an example, the smart car can only exist with infrastructure with which it exchanges transparent information.

3.3.1.4. *Technical assistance*

First of all, helping systems have the ability to really assist humans and to facilitate their production tasks. This process requires aggregating, processing and visualizing information to make it intelligible. This helps people to make decisions "progressively" and to solve urgent problems based on this information. As well, the capacity of cyber-physical systems makes it possible to physically help humans. They perform a variety of tasks that are tough, uncomfortable, exhausting, unsafe or sometimes dangerous. Robots perform part of these technical assistance actions.

3.3.2. *The place of digital*

"Digital" is the main factor. Its variations are multiple. Digital technology makes it possible to create three-dimensional plans, to carry out simulations, to use sensors, to enjoy artificial vision modes, to enable online tests and, above all, to communicate easily across the planet. Besides, it eases the management, not only of production and the entire supply chain, but also of customers and the relationships with administration. Digitization is also a key point in the simulation and control of industrial processes, which have become automated. Factory employees only have three types of tasks: machine monitoring, rounding up and quality control, although the latter is partly automated.

The places and consequences of digital technology are represented in Table 3.2.

	Connected and computerized objects and tools	**Big data**	**Augmented and virtual reality, artificial intelligence, recognition and segmentation algorithm**
Production automation	Decrease in the price of sensors and multiplying of their possible applications Boom of calculating and storing capabilities (cloud)		
Productive maintenance	↓		
Remote control			
Production tracking and marking			
Input consumption efficiency (energy, raw matter)	Productivity gains Quality improvement Tailor-made production of small amounts Development of "servitization"		

Table 3.2. *Model for industry 4.0*

3.3.3. *Direct manufacturing*

The possibility of "direct manufacturing" is an important determinant, whose elements are known as "3D printing", "continuous extrusion", "numerical control programmable machines", "automatisms" and "programmable robots".

In society 5.0, this notion is not central, but decisive. Additive manufacturing has always existed, and we can confirm it by observing the work of potters, who are used to working within this framework. The important point to bear in mind is the possibility of its generalization, and therefore, of being able to produce closer to the consumer.

Direct manufacturing includes additive manufacturing, but these two should not be confused.

3.4. The different types of factories of the future

Our research has enabled us to identify five types of factories of the future:

– factory 4.0: integrated logistics chain, often presented as the German model;

– the Key-Technology factory, based on a highly differentiating process;

– the Craft-Industrial factory, in which "tailor-made" production becomes industrialized;

– the Client Drive factory, where the customer launches the process;

– the Low Cost factory, whose elements are most often operated in Open Source and are frequently self-propagating[2].

In the following sections, we will try to describe their regulatory determinants. One of the key determinants of this new type of industry is the shift from an industry dominated by product-analysis to a customer-service-oriented process industry. The different types of factories of the future will be developed in the lines that follow.

2 For example, 3D printers have been conceived in such a way that they can be built with an identical 3D printer, plus the addition of some elements.

3.4.1. *Factory 4.0: "integrated logistics chain"*

DEFINITION. The factory works as an "integrated logistics chain", integrating different elements in close relation with a multitude of suppliers.

Described as the "German model", this factory is present in the automotive industry. The Renault tractor plant at Le Mans, now turned into Claas, is the best-known example. Set up in the 1970s, it made it possible to manufacture all types of tractors, from the smallest to the largest, as well as military equipment on the same assembly line.

3.4.2. *The Key-Technology factory: "a highly differentiating process"*

DEFINITION. The Key-Technology factory is based on a machine-centric process equipped with highly differentiating technology.

For example: the central element of a reactor blade is produced thanks to the continuous development of a huge crystal.

3.4.3. *The Craft-Industrial factory: "tailor-made industrialized production"*

DEFINITION. The Craft-Industrial factory makes it possible to generate tailor-made industrialized production.

For example: instead of hand carving a block of marble to build the "sculpture" destined to the top of the wedding cake of newlyweds, a craftsman works with a panoramic camera and a 3D printer. Let us also mention the craftsman who uses a 3D printer for the after-sales service of broken pieces of household appliances, and this, directly following the plans from the manufacturer. Robots and Computer Numerical Control (CNC) machines are the key to this type of industry.

3.4.4. *The Client Drive factory: "the customer operates the process"*

DEFINITION. In the case of the Client Drive factory, it is the customer who operates the manufacturing process.

For example: a customer is "virtually" trying on different pairs of glasses with various frames. The payment of his basket triggers the production process and then, the shipping of his glasses by the manufacturer. In another example, a customer automatically takes his measurements thanks to a double camera on his PC, chooses his fabrics and his colors. The payment of his basket triggers the production of his three-piece suit. The same would apply to the case of the customer configuring his car on the Internet.

The automobile industry had this objective in mind in the 1970s, when it computerized its production, as was the case of the Volvo plant in Sweden. This industry was then put on hold, before returning more recently, for high-end vehicles. The decisive point is the cost of the investment needed in information technology, so as to move from customer demand to production. France Telecom and Deutsche Telekom had relied on this type of tool in the 1990s in order to build digital telephone lines. Many Client Drive factories are an upgraded version of the "integrated logistics chain" type of factories, but this is not always the case.

3.4.5. *The Low Cost factory: "in Open Source"*

The Low Cost factory stems from the expanded Open Source. Crowdfunding has become the central element of market research, in terms of funding, planning and issuing the first pre-orders. Blueprints for 3D printing on certain websites are royalty-free and editable. These elements are directly introduced into an in-Moove robot, and they drive the sensors. Programs for Arduino or Raspberry also exist. In this case, it is this equipment that controls the sensors. The importance of this approach is that it makes it possible to create localized or home-made production lines, something we call a desktop factory. This type of personal fabrication has two characteristics:

– it is a type of manufacturing that has no design costs and therefore imposes the existence of free design or the need to find a way to pay for this design;

– it is a type of production more likely to match the expectations of the potential user.

The concept of the Low Cost factory has been reused by large manufacturers in order to place their production closer to consumers.

Trakett, a manufacturer of laminate flooring, is one example of this approach.

The self-propagation associated with the Low Cost factory is rather a gadget concept. In fact, it either applies only to "has been" machines, or to machines whose performance has not yet been optimized. This is why the impact of self-propagation remains limited for industry 4.0.

3.5. The regulatory determinants of the factory of the future

The problem of value is complex in society 5.0, but actually this situation is not new. Two centuries ago, Henry Ford stated: "If I had asked people what they wanted, they would have said faster horses". The creation of the new society 5.0 imposes disruptions, which are difficult to imagine with conventional methods. In other words, incremental methods must be questioned, as is the case of classic lean manufacturing. The standard is no longer continuous improvement, but switching, associated with breaks and disruption.

3.6. The main questions regarding the factory of the future

The determinants of the factory of the future are multiple and can be summarized by a few questions described in Table 3.3. All these questions will have to be studied and pondered in the future.

Questions	Observations
Who will be the creator of the factory of the future?	The current fashionable start-up model only builds a few factories, but generates platforms and services that use products built elsewhere
What will be its origins (political will, individual actions, etc.)?	Individual actions and big trends like bio and the entertainment society seem to be its driving forces
With whom, which capital sources and which capital distribution?	The cooperative mode is being reborn
For whom, who will be the main beneficiaries?	The consumer seems to be the commonly agreed reply, but surely a false one
Who will lead it or be responsible for it?	The question aims at characterizing the educational background and skills of the leader

Questions	Observations
What will be the different types of actors of the factory of the future?	This invites us to think about new occupations and their matching skills
What will be the subjects, targets and products and services aimed?	The industry of the future will affect some sectors more than others: those which are already well computerized or, on the contrary, sectors which will make a big leap
What will the factory of the future be like from the "physical point of view"? What will it be made of, with what materials (building), related to what?	As we have shown, the factory of the future will be nearer consumers, therefore less massive. It is evident that traditional buildings and uneven production will not reflect the architectural organization of the factory of the future
The energy-based factory of the Industrial Revolution will be replaced by other industrial forms that need to be described. Which tools will be involved in this factory?	This question is important because it concerns the means that will have to be added to those of information, communication and transportation
When will it become effective?	The question of performance is complex itself
How will the factory of the future work?	The notions of non-stop factory, 3 × 8, etc. are no longer relevant
Under what conditions, how will it come about?	The question here is that of constraints
Apart from digitization and greener manufacturing, what other processes will be at the heart of the creation of the factories of the future?	The share of sustainability must be balanced with that of cost reduction
What will be the correct and necessary procedures?	The operating rules will surely have to be negotiated, particularly with the unions
Which techniques and material resources will it need to implement?	The question of machine tools and their distribution should not be overlooked

Table 3.3. *Determinants of the factory of the future*

A new theory on industrial economy will have to be built. What are the main elements that will describe it? This is a philosophical revolution in which the distinction between "capital and labor" disappears, and new descriptors need to be considered. The factory of the future will be based on a vision that determines an object based on goals, expecting results, either in financial or in productive terms.

3.6.1. *The location of the factory of the future*

QUESTION. Where will the factory of the future be placed?

A few years ago, production was concentrated in places of energy sources or endowed with raw materials. Metallurgical plants were born in the mining basins. They later moved near the ports in order to benefit from the arrival of raw materials. Other factories settled near their customers. Others settled near workforce sources associated with the peasantry, actually in the countryside. The questions: "where? from where? in what direction?" should not be overlooked. They are key factors and a guarantee of future success. The location of the factory of the future is an essential point. The only certainty is that many current locations will be abandoned.

A chapter of this book will focus on these location issues.

3.6.2. *Production cycles*

The creative phase and the cycles of the factory of the future also need to be described.

QUESTION. When and why create a factory?

The notion of *kronos* now becomes relevant.

A reflection on the determinants of production cycles, traditionally called "seasons", needs to be encouraged. Seasons used to be associated with modes. Some clothing manufacturers have revolutionized production and distribution by abandoning the two traditional fashion seasons (summer and winter), and making the best use of sales and customer data[3]. How long does a production cycle or a collection last? Key dates, periodicity and duration

3 We are making reference to Inditex, whose key brand is Zara, and Desigual, both of them, Spanish.

are significant elements. In the electronics field, the IFA[4] held in Berlin every September, the November Comdex[5] in Las Vegas and the Hanover Cebit[6] held in April are the three key meeting dates for producers of electronic devices. Is it necessary to point out that none of these large salons is held in France and that, moreover, giant events of this type are being created in Asia? For example, the HKTDC Hong Kong Electronics Fair is held twice a year in Hong Kong and, including its satellite events, it hosts more than 5,000 suppliers.

3.6.3. *Finances in the factory of the future*

The factory of the future also raises financial questions. These primarily involve cost and investment aspects.

Questions	Observations
How much will it cost?	The standard cost of a current factory is calculated in millions, and in several billions for semiconductor factories. A good initiative could be to lower production costs
What will determine financing sources?	The current financing of start-ups does not work for the factory of the future and the question that currently arises is on how financing should be oriented
How will investments pay off?	It is necessary to consider the depreciation time of transformation

Table 3.4. *Financing the factory of the future*

Performance issues and its indicators should not be left aside. Nevertheless, if material balances and production costs are down, this will not be enough. It is important to understand how the factory will replace traditional solutions, in which economic sectors and for which reasons. This is the question that will be explored in the rest of the book.

---

4 *Internationale FunkAusstellung*, which literally means International Communications Fair.

5 COMDEX is the short form for Computer Dealers' Exhibition.

6 CeBIT is the German acronym for *Centrum für Büroautomation, Informationstechnologie und Telekommunikation*, which can be translated as "Center for Office Automation, Information Technology and Telecommunications".

3.6.4. *The conditions of its emergence*

The arrival of the factory of the future and the disappearance of previous actors also raise some still-indefinite issues.

Questions	Observations
Why will the factory of the future be capable of eliminating more traditional production means?	The place of the factory of the future in Porter's strategic layout
What will be the causes, the facilitators, the triggering factors, the motivations?	Identifying key points and the gear shift locks
Which risks will be run?	A risk analysis is necessary, but should not put the project on hold

Table 3.5. *The conditions of its emergence*

In a similar way to previous situations, we should not forget that, as generally happens with any new advance, false explanations, beliefs and myths are to be expected for the factory of the future. Detractors and opponents of all kinds will not fail to appear.

As far as human resources are concerned, the competence and training of people will be decisive points.

Motivation, purpose, goal, justification and the question of why are central. The expectations, wishes, ambitions, forecasts and expected results will only make their appearance, provided that certain elements are respected and that some constraints are reduced. We will try to explain them. What kind of factories will be associated with certain types of products?

3.7. Changes related to the factory of the future

The transition from the product to different types of services is the decisive economic factor. "Servuction", "servitization" and "servicing" become terms that are important and represent new ways of marketing. If these elements are certain, a new question will have to be solved.

QUESTION. How will the margin be built and what will the pricing strategies be?

3.7.1. *Actions for favoring the advent of the factory of the future*

Several types of actions are possible:

– "pull" actions involve calls for expressions of interest (every sector) or calls for projects inspired in an industrial theme (foundry, prosthetics, ceramics, etc.);

– "push" actions are multiple. These involve the creation of Fab Labs, the action targeted at young people (high school) or university students (thematic plant nursery) and the establishment of seminars devoted to cost-saving technology (Arduino micro-computer, sensors, Low Cost mechatronics). The establishment of technical centers (sensor) and Fab Add clusters are significant incentives. Finally, regarding communication to university teams, policies may be interpreted as a guarantee of success.

3.7.2. *The notion of industrial revolution*

Will the factory of the future represent an industrial revolution in the sense of Rifkin[7]?

DEFINITION. An industrial revolution is a period linked to evolution in energy and communication modes.

QUESTION. If we use this definition, what ethics should be adopted, and what might be the social and societal tensions that could come with the new type of factory?

In order to avoid the term "revolution", many actors, including politicians, have used the term "transition", which can be found in formulas such as "digital transition" or "ecological transition". This approach has the disadvantage of presenting some "holes", in that it fails to make the link between the digital transition and the ecological transition.

7 This is a reference to Jeremy Rifkin's book, *La Troisième Révolution industrielle. Comment le pouvoir latéral va transformer l'énergie, l'économie et le monde*, Les Liens qui libèrent, Paris, 2012.

3.8. Daily management

The questions regarding the management of the factory of the future are as follows:

QUESTION. How will the factory of the future be managed on a daily basis?

This poses different sub-questions, including the question of resources. People and labor relations, as well as their place in the factory, are factors not to be neglected in the understanding of the factory of the future.

Tool manufacturing, especially the specific machines that manufacture other machines, inventory management or the notion of PoiCo all disappear with automation and tight flows. The concept of stock will evolve, while the notion of PoiCo (point of command) will probably be left aside in favor of Stock Keeping Units (SKU). This is how produced units are referenced in a Global Trade Item Number (GTIN). The Universal Product Code (UPC), European Article Numbering (EAN) code or the Australian Product Number (APN) are all special cases.

3.9. Additive manufacturing technologies

Manufacturing technologies have evolved a lot, due to the rise of digitally controlled equipment.

3.9.1. *CNC tools*

CNC tools are the predecessors of more complete and standardized systems. Nowadays, these tools have become widespread.

3.9.2. *The notion of CPPS*

One of the new paradigms of industry 4.0 is the Cyber-Physical Production System (CPPS).

This concept dates back to the early 2000s and was defined by the National Science Foundation (NSF) in the USA.

DEFINITION. A Cyber-Physical Production System (CPPS) is a system in which IT and telecommunication elements collaborate to control physical entities, particularly in factories, but also in the distribution of goods and services.

3.10. The example of the textile industry

In order to launch a new textile product, work starts on a computer to prepare the patterns and the bill of material. Each product is then coded. Then, every part is printed in real size with the various indications needed for its assembly. At the same time, a three-dimensional shape appears on computer screens. At this stage, a process for managing the product's lifecycle begins. Lectra[8], a French firm, is one of the world leaders in this process. As this is a centralized information-sharing platform, the software offered by this firm ensures the link between the planning, creation, development and sourcing teams. At this point, we can understand the importance of two concepts of society 5.0, such as information exchange (*datum*) and that of platform, which we will address later. This platform makes it possible to visualize not only the collection plan, but also its associated production process, the key dates of the schedule, the assortment plans and their logistical consequences. It favors optimal control, at the macro- and micro-economic level, at the time that it offers cost reduction tools. This reveals the interest of data visualization, which we will discuss later. With this type of software, firms can better manage resources (input), schedules (dates) and costs. It is also useful for formalizing relationships with external suppliers.

In order to add value to ideas as quickly as possible, software offers professional aid for creation, so that these ideas can be rapidly transformed into marketable textile products. These solutions are based on intelligent processes. They enable the creative team to focus solely on pure creation, and free them from other, automatable, and often non-value-added tasks.

There are two distinct stages: the choice of the "textile" and the design of the model and its alternatives.

8 For more information: http://www.lectra.com.

The system makes it possible to create shapes, to choose printed, woven or mesh textile as well as to manage colors. In a first phase, prototyping and its iterations make it possible to verify that the product meets the customer's expectations of the model. Besides, this prototyping is carried out using simulators that can yield a 3D rendering.

The production phase begins by cutting the pieces of each product in the selected fabric or tissue. Software has made it possible to determine a way of cutting fabric with an optimality function that makes it possible to minimize losses as much as possible. In other cases, software can optimize the mesh shapes when simultaneously developing shapes and knitting mesh fabric. This procedure is often called a cutting plan and may be several meters long.

Then, a calculation is performed so as to cut out the largest possible number of tissue layers in one go. This number depends on materials, their thickness, their grammage and the cutting tools employed (saw cutting, laser cutting and water cutting under pressure.) The cutting of the fabric is performed according to the cutting plane. These cutouts are being performed by increasingly automated systems called "sewbots". A laser offers multiple solutions and functions for cutting materials, since it cannot only cut, but carve and perforate. As a technique, laser cutting is sharper than other solutions. Omega Systems, a French firm, offers this type of tool. The laser cutting process is managed by a computer that determines not only the power and speed of the laser, but also its potency. This makes it possible to adapt the cutting depth according to the material worked, as well as the quality of the cut. Engraving, automated holes and marking are also possible thanks to the accuracy of its settings. This saves great amounts of time in identifying pieces or in making buttonholes. This cutting mode not only avoids fraying (by cauterizing the edges), but it also eliminates overcasting (the following task), something we will discuss later on. Many materials are laser cut, which is not the case with other technologies. Omega Systems provides a list of the following materials: "textile, foam, leather, medium, polyesters, M0, wood veneer, soft floor, polycarbonates, environmental fabrics, baking consumables for composites, plywood, PES polyester, airbag, printed matter, film". As a result, creativity is increased tenfold and makes it possible to conceive products that used to be very complex. The water jet cutter technique is also very much in vogue. The use of 3D technologies with cutting angles makes it possible to cut several sizes of clothes in the same stack of fabric.

Reusable fabric pieces, minimized thanks to the computerized solution, are recovered for other uses, such as padding or tissue sampling for future customers. According to this cutting mode, fabric pieces are characterized by raw edges that fray. When necessary, intelligent overlockers locate the edges of fabric and overcast them with zigzag stitches, with an edge cut to 1 or 2 millimeters. With laser cutouts, this action is sometimes not even necessary.

The pieces of fabric are sent to seamers to be assembled. While work is often manual at this point, sewing robots are starting to be implemented. Task distribution is set up and software helps to organize distribution in such a way that no time is wasted.

Once assembly is complete, quality control is the following step. Specialized firms introduce their logos and brands, and verify possible labels and barcodes. Other employees handle the packaging, price marking and shipping logistics.

4

The City and Mobility 3.0

Society 5.0 is defined within cities 3.0 or villages 3.0, in parallel with mobility 3.0, which can be defined as "smart", or even "smarter". These notions are complex and do not necessarily represent what ordinary people infer from this approach. For others, especially for the Japanese, these are concepts that are no longer hot topics. The city and mobility are the key elements of this "revolution". As a result, the construction and public works industries (namely the automobile industry) will be upset. Cities 3.0 are often called "smart cities", which is often a misnomer.

4.1. Research

While in other sectors researchers have not been interested in the field of research or have spoken without knowing the terrain, this is not the case of the city and mobility, which have been the subject of deep original research. Our book gives some examples of this. While there are some research actions focused on the city, others focus on vehicles.

4.1.1. *The city in motion*

The notion of a moving city stems from the idea of the influence of the city on the rural world. As Francis Godard observed in his book:

> "In every industrialized country, the urban population is nowadays larger than the rural population, and the metropolises in developing countries have known and are still experiencing vertiginous growth. The homogeneous and compact city, with a

> city center and well-defined boundaries, belongs to European history; conurbations, gigapolis and other megacities are announcing the advent of a new urban civilization"[1].

Sharing the same approach, François Ascher is a French urban planner and sociologist who specializes in the study of metropolitan phenomena and urban planning. He coined the terms "metapolis" and "hypermodernity", and was president of the Institute of the City in Motion.

With the concept of hypermodernity, the author made a connection between the contemporary individual and his metropolitan condition, as well as his displacements. He also called into question the possible coexistence of constructed unique individualities and multiple urbanities. The concept of metapolis refers to these compact, loose and discontinuous urban spaces. These spaces continue spreading and are inseparable from the metropolitan condition of individuals, which compels us to keep moving.

The Institute for the City in Motion (*Institut pour la ville en mouvement* or IVM) is an international organization that conducts research and innovative actions in the field of urban mobility. Created in June 2000 thanks to the financing of the PSA-Peugeot Citroën group, it has long been directed by François Ascher. This establishment has offices in Buenos Aires, Paris, São Paulo and Shanghai. Since January 2016, IVM has become a member of VEDECOME, an organization for energy transition (promoting carbon-free vehicles and their mobility). This is a public-private partnership research and training institute, dedicated to autonomous vehicles and low-carbon and sustainable individual mobility.

4.1.2. *Transit-City program*

Transit-City[2] is a program for reflecting upon the future city and lifestyles. The program aims to build bridges between areas as diverse as urban planning and marketing, retail and sociology, and real estate development and transportation. It mainly focuses on city 3.0 and mobility 3.0. This program was born from the desire to decompartmentalize

---

1 Translation of French quote. We are referring to the book by Francis Godard, *La Ville en mouvement*, Gallimard, Paris, 2001.

2 For more information: http://www.transit-city.com/.

prospective reflection by regularly bringing together urban actors coming from very different horizons (real estate, transport, trade, architecture, design, history of the art, etc.), who do not often get the chance to meet and exchange their viewpoints. This inspired the idea of conducting workshops and observatories.

For this, researchers defined the concept of transit universe, which includes the commuting spaces of individuals in a fixed data system, such as infrastructure and housing facilities. In this way, the observation of universes is interesting for understanding the place of the city, of mobility 3.0 and of industry 4.0 in society 5.0.

4.1.3. *Research on smart vehicles*

The search for an autonomous vehicle is not new. Cars, in particular, have been affected by this research only more recently.

The autonomous aircraft and the aircraft's electrically controlled automated piloting emerged thanks to a project called "the cockpit of the future", implemented by the firm Altran in the 1980s. More recently, the fully electric Solar Impulse aircraft made the world tour. In the 1990s, Altran developed the embedded local network of TGVs.

The *Villeneuve-d'Ascq-Lille* (VAL) project is a fully automatic metro run on tires, destined for urban transport. The acronym VAL has been transformed into "light automatic vehicle" ("*véhicule automatique léger*" or VAL). This project is one of the first of its kind in the world. It originated in the research of university laboratories and the Industrial Institute of the North. The patent concerning the automatisms of a driverless metro was filed on July 31, 1971, by Professor Robert Gabillard. Originally built by Matra, this firm has now become Siemens Mobility.

Regarding the car, the invention of the lead-acid car battery by Gaston Planté in 1859, and later on, the works of Camille Faure, made the development of the electric car possible. In 1899 in Belgium, "*La Jamais Contente*" was the first electric car to exceed 100 km/h. In 1900, electric cars represented more than a third of the world's fleet. A radio-controlled car without a driver circulated in New York in 1925 and was officially introduced at the New York International Fair of 1939. The electric car was relaunched with oil shocks. In 1972, General Motors (GM) built the first hybrid car, the Buick Skylark. Victor Wouk is often referred to as the godfather of the hybrid vehicle. In 1974, the Vanguard-Sebring CitiCar made its appearance at the Washington Electric Vehicle Symposium, resembling European unlicensed vehicles.

The European project Promothéus, which stands for Program for a European Traffic of Higher Efficiency and Unprecedented Safety, intends to encourage a revival of the idea. The aim is to improve long-distance road traffic, both for the vehicle and also regarding the infrastructure. The main manufacturer who attended the event was Mercedes-Benz. From this came the idea of the end to "traffic jams". The project, which started in 1987, produced several prototypes in the 1990s. Some of these were abandoned (as was the case of Renault). Others, like the one of Mercedes-Benz, circulated and made a journey under normal conditions between Munich and Copenhagen. The main points raised by these projects were:

– the need for a driver to regain control every 10 km, on average;

– the difficulty for implementing safety strategies, especially in agglomerations;

– the gain that a vehicle would provide to the engine and to the electric controls (as was the case for the aircraft's orders).

Volvo proposed safety systems for low-speed urban situations, including automatic braking in 2008. Google started testing a self-driving vehicle in 2009, based on map learning. In 2013, some manufacturers' models were equipped with automatic braking, encounter avoidance, persistence within the road boundary lines, automated parking, adaptation of vehicle speed to traffic and automatic reading of some road signs. The firm Tesla announced its electric autopilot in 2014. An accident with a Tesla autonomous car hit the newspaper headlines.

4.2. The link between smart vehicles and road infrastructure[3]

In the history of transport, the progress of the automobile has been associated with changes in infrastructure, at the time that tarmac roads have made life easier for tires, explaining the abandonment of cobblestones. The

---

3 Some elements are drawn from Leonard's study (Vinci's Open Innovation Department) commissioned to Fabernovel in 2017.

same applies to the vehicles of the future. The link between vehicles and infrastructure becomes a determining factor. The existence of this link will require significant investment, which will have to be borne by States and infrastructure planners, or be integrated into the "servitization" of people or merchandise transport.

Thanks to its numerous sensors, the new vehicle is able to collect data relating to its cabin and its external environment. Besides, it is connected. By means of its various forms of connectivity, it offers the possibility of new interaction possibilities with occupants and infrastructures. This is why the reflections on the vehicle-infrastructure couple are the salient feature of the upcoming society 5.0. The problem originates in the unfortunately unclear vision that automakers and automotive suppliers have concerning the connected vehicle of the future. In addition to this perspective problem, we can mention lack of coordination for defining standards. Various issues currently remain unsolved and are described in Table 4.1.

Questions	Observations
What kind of technology will grant uninterrupted connectivity?	This is one of the challenges in the development of new mobile phone generations, particularly the fifth generation (5G)
How to adapt to the context?	All current computer technologies are based on context-free language. It will be necessary either to invent context-related language, or to simulate this with current languages
Which are the different functions and the associated infrastructure?	These topics are complex, as for other technologies. The use of a mobile phone is not the use of an old phone, which has become mobile.
Which are the necessary applications and functions dedicated to users?	Users have difficulty defining needs. Living labs can be useful. The new features will appear in the form of improperly used functions.
What are the sources of value for the users and the field actors?	This problem is fundamental in the market approach

Table 4.1. *Unsolved questions regarding the smart vehicle*

4.2.1. *Smart vehicles' levels*

The development of autonomous vehicles has led authorities to create standards for classifying cars according to their level of autonomy. In fact, there are two scales: the American one (0 to 4) and the European one

(0 to 5). The European scale is more precise and splits up the third level into two sub-levels. The classification was built according to the number of primary functions supported by the on-board computer.

Level	Function	Observation
0	*Single driver:* keeps complete control over all vehicle functions: brakes, acceleration, steering	The onboard computer can assist the driver via alert messages when the car speeds up or crosses a line, or it can display images using a reversing camera
1	*Assisted driver*: the onboard computer can handle speed and/or direction, while the driver keeps control over other functions and total control of the vehicle	Speed control can be adapted according to preceding vehicles
2	*Supervising driver:* partial automation where the vehicle controls speed and direction, particularly for parking inside or outside a garage	This is temporary driving of the vehicle. The driver merely oversees the operations. He must nonetheless be very attentive and take back control in case of system failure, because his responsibility is always fully engaged
3	*Conditioned automation*: the driver totally delegates the driving process in pre-established situations	The onboard computer knows its limitations, and the "driver" must nonetheless be capable of recovering full control
4	*Absence of driver in certain situations:* in a limited context and a predefined situation, the car is capable of moving without its owner	The two most common examples are motorway driving with specific infrastructure and the example of a car which is capable of parking in a parking space. The car can also pick up its owner when necessary
5	*Completely autonomous car*: the onboard computer controls all the car functions	One of the greatest complexities for the onboard computer will be able to consider and to quickly analyze the reactions of humans who are still driving

Table 4.2. *Autonomous car levels*

These levels refer to the industrial reality and to vehicles which are already in operation.

4.2.2. *Current examples of autonomous vehicles*

Fully autonomous cars may never truly arrive. Projects like the European Autopilot Project intend to improve autonomous vehicles thanks to Big Data. On the other hand, Intel is showing how autonomous vehicles will communicate in the case of unforeseen events, especially in accidents that paralyze traffic or during work detours. PSA is working on this subject: on July 12, 2017, for the first time, a Citroën autonomous vehicle passed the barrier of the largest highway toll in Europe "on its own". The growth of more autonomous vehicles will involve real-time communication between the road and the vehicle. Tesla's Autopilot corresponds to level 2, while Audi is announcing a "level 3" autonomous vehicle including "eyes on, hands off". The "traffic jam" function of the Volvo XC90 is included in this category. Daimler and Google are working on an autonomous electrical car, without a pedal or steering wheel, and adapted to carpooling.

4.2.3. *The challenges of the road environment*

The road environment will have to be connected. In fact, the vehicle which is becoming more and more autonomous will have to benefit from constant communication between the vehicle and the infrastructure. Fifth-generation mobile telephony (5G) should provide two layers: one devoted to passenger services in the vehicle, and the other to make exchanges between connected objects. By freeing himself from the driving function, the driver will become almost a passenger like the others.

The scenarios for means of transport use will also evolve. The public transport user will no longer have to own a car. They will call a vehicle with their smartphone; the vehicle will arrive at the place where the passengers are and will drive its passengers to their destination. The user will only have to pay for the traveled distance, the quality of the vehicle or time used. Commuting time can be productive, the drivers can profit to entertain themselves or to get to know the places they will go by, informed for marketing purposes (for example, a restaurant) or for cultural purposes (a monument to visit). Vehicles have moved from the driver's stage to the democratization era, thanks to the appearance of the driving license and driving by oneself. Future generations, our children, will never need a driver's license. These generations of users will no longer buy cars. The users will no longer need to park their vehicles. City parking policies will have to be reconsidered. All this will not only transform the cities, but also the countryside. We will no longer need vehicles or car parks. Users will need transportation infrastructure.

4.2.4. *The smart and mobile habitat*

The smart habitat has also been examined in numerous studies, whose main types are introduced in Table 4.3.

Name	Type of observation	Action
Catastrophic Cities	Sustainability and the city	Program of reflections on the possible consequences of natural disasters, on how to ponder them, how to design cities and their buildings, and how to manage their mobility
Scale(s)	Distribution of products and services	Cell watch and prospection of trade developments
Fiction Cities	Cities and urban areas	Resource center on the role of fiction in the construction of urban imaginaries
Habitat(s)	Habitat	Cell watch and prospection of the evolution of habitats and ways of living
Next Cities (City of the future)	Cities and urban areas	Resource Center on Urban History
Next Factory	Industry 4.0	Observatory on the evolution of the world of work, its methods and its spaces
Transit(s)	Mobility 3.0 (Nomadism)	Sociomarketing watch cell to assess the influence of mobility on our lifestyles and the creation of new products and services

Table 4.3. *Type of analysis performed on mobility 3.0, inspired in the work of François Bellanger*

5

Information Technology 2.0, the Foundation of Society 5.0

Digital transformation is a concept that takes up many different forms and is involved in all aspects of human life. Wasterman (2011), Stolterman and Fors (2004) are the authors who most often bring these topics forward. It is necessary to analyze the situation created by the existence of a digital man. At the same time, we must be aware that information technologies are the foundation of society 5.0 which has led to a triple phenomenon characterized by more freedom, more control and the importance of communities.

5.1. The reference to Jean-Paul Sartre

The current situation was described by philosopher Jean-Paul Sartre far earlier than the advent of the digital era:

> "Freedom is one, but manifests itself differently depending on the circumstances. To all the philosophers who defend it, it is permissible to ask a preliminary question: regarding which privileged situation have you experienced your freedom? One thing, in fact, is to feel that one is free in terms of action, of social or political enterprise, of creation in the arts, and a different thing is to experience it in the act of understanding and discovering"[1].

1 Translation of French quote. Jean-Paul Sartre, *Situations philosophiques*, p. 61, Gallimard, Paris, 1990.

One of the important contributions of existentialism, a characteristic of Jean-Paul Sartre's thought, is defined by the principle of human freedom. Sartre refused to think outside of factual situations where freedom is exerted. For this, he took into account the place as well as the period. People do not wish to relinquish their freedom. The existence of man makes him seek to escape from any form of limitation. Therefore, we can perceive the importance of the digital world in the search for freedom and the disappearance of boundaries, barriers and borders. Citizens involved in such a quest for information, to the point that they seek it at the other side of the world, on a server. People can communicate with others through these tools, even if society prevents them from doing so.

When Sartre stated that "man is condemned to be free", he recalled that freedom is not a foundation in itself.

In a well-known section of *L'Être et le Néant* (Being and Nothingness), Jean-Paul Sartre explained the reason why freedom is not natural:

> "But if freedom is to be its own foundation, then the end must in addition turn back on its existence and cause it to arise. The possibility of being free and the possibility of not being free also exist before the free choice of freedom. But since it might be necessary to have prior freedom in order to choose to be free (basically freedom choosing to be what it already is) we might be sent back to infinity because it might be necessary to have another previous freedom to choose it and so on. In fact, we are a freedom that chooses, but we do not choose to be free: we are condemned to freedom"[2].

Sartre is highly decisive for analyzing our societies. He is interested in the condition or the situation of man. This is determined by a set of relative data, representing spatial, temporal and societal situations, as included in a group. In this way, we can understand the importance of smartphones, for which localization is important data in their general use, as well as for specific functions of social networks, such as Facebook. These social networks make it possible to place ourselves within groups and within the temporal evolution of an event via the use of comments.

---

2 Translation of French quote. Jean-Paul Sartre, *L'Être et le Néant*, p. 530, Gallimard, Paris, 1943.

Despite the fact that the position of man is located, it does not work as a restriction. In fact, this is one of the key elements of Sartre's thought:

> "We shall use the term *situation* for the contingency of freedom in the *plenum* of being of the world inasmuch as this *datum*, which is there only *in order not to constrain* freedom, is revealed to this freedom only as *already illuminated* by the end which freedom chooses" [3].

Thus, for Sartre, no element in the world is in itself an obstacle or an auxiliary, a promise or a threat. Therefore, this approach should be linked to the strategic method which analyzes strengths, weaknesses, threats and opportunities. It is quite interesting to observe that the consultant or the strategy auditor begins with the weaknesses and ends with the opportunities, which introduces certain dynamics that are at odds with Sartrian thought.

Sartre focuses greatly on resistance and very little to facilitators, which is characteristic of tools built by man. One of the goals of setting up society 5.0 is to focus more on facilitators rather than on resistance:

> "Thus we can begin to catch a glimpse of the paradox of freedom: there is freedom only in a *situation*, and there is a situation only through freedom. Human-reality everywhere encounters resistance and obstacles which it has not created; but these resistances and obstacles have meaning only in and through the free choice that human-reality *is*" [4].

This situation is actually a kind of synthesis of different data which digital actors seek to capture by means of platforms and applications. It is a question of me and my body, my place, those around me, my past and my desire for the future, my belongings, etc.

5.2. The "Sartrian" man in the digital world

The Sartrian man is defined by different needs. Some of his needs are internal and appear as strengths, but also as threats: these are self-knowledge and self-perception. Other needs rather appear as weaknesses and

---

3 Translation of French quote. Jean-Paul Sartre, *L'Être et le Néant*, *op. cit.*, p. 532.

4 Translation of French quote. Jean-Paul Sartre, *L'Être et le Néant*, *op. cit.*, p. 534.

opportunities, such as the knowledge of others and of the outside world. Digital technology allows us to respond to the need for this knowledge, provided that we know how to use the tools of the digital world. Let us then refer to a sentence attributed to Sun Tzu:

> "If you know yourself and you know your opponent, you will win all wars".

The digital man, or *homo numericus,* is a person, but does not represent all aspects of this digital world, because, as Sartre taught us, the digital man is bound to a situation.

Digital men gather spontaneously in digital communities, which represent one element of the situation in Sartre's sense. But these are not all the elements that can help us describe the situation. That is why research work is needed so as to understand other complementary elements.

A strategic analysis of the situation of man in the digital world is summarized in Table 5.1.

Strengths	Weaknesses	Threats	Opportunities
Access to information Easier technological education and training	Informational overload (psychological effect) Infobesity Boundary disappearance and reduction of the influence of Nation-States Much (too much) time spent on media	Information manipulation for propaganda purposes Respect of private life Disappointment due to the difference between the real world and the virtual world End of certain types of property Restrictions Information versus knowledge Reproduction facility	Opening to new worlds Advanced effect of digital on sectors like agriculture and health Making educating citizens easier Possibility of building objects at monumental scales End of an old world

Table 5.1. *SWOT analysis*

The digital world enables us to remain connected, not only virtually but also in the real world. It is interesting to reckon the influence of the digital world in revolutions, especially in countries like Tunisia or Egypt, where it helped to remove the communication barriers imposed by dictators.

5.3. Schemata

The notion of schema[5] is one of the fundamental points of the construction of society 5.0. Many difficulties affect the collection, storage and research of data. A single schema is made up of several data concerning an individual or a person. Mathematically, a schema can be represented by a vector.

Schemas as meaningful data, in this case, are providers of knowledge as well as sources of skills. Sometimes, schemas are the identification data. They make it possible to be located in time, in space or in both. The schemas can be simulation or calculation data. It is for this reason that it is always interesting to separate the three types of data in the analysis that we perform on data in society 5.0. Finally, society 5.0 has created a new form of data: that of commands, which have no other use than the fact that they make it possible to control the elements (including machines) of society 5.0. This notion of schema possesses a higher conceptual level than the notion of information code and signal information described previously. These two types of information are nonetheless the ones that make it possible to build schemas.

5.4. Data in their environment

As we have previously seen, a situation in society 5.0 is strictly linked to the context, which influences the types of data.

5.4.1. *The sources of data*

Around 20% of people in a Western country are spied on via their smartphone in all of their activities. This is achieved thanks to downloaded

5 For Kant, a schema is an intermediary representation between phenomena perceived by the senses and the categories of understanding. Jean-Paul Sartre preferred to use the term *datum* for this.

applications and marketing software. This figure reaches 80% if we consider basic data transmitted on the network.

When you eat at a restaurant, someone knows it thanks to your smartphone, your credit card or your loyalty card. The institution will sell this piece of information to commercial firms who want to know more about you. The behaviors of users scrutinized in this way are later analyzed by schemas including algorithms and associated means of data storage (Big Data). To this, we have to add geolocation data.

The firm Teemo[6] (formerly Databerries) is a key player that acts through commercial applications from partners such as Carrefour, Leaderprice, Leclerc, Monoprix or through media such as L'Équipe, Le Figaro or major brands, such as Volkswagen. These firms reached an agreement with Teemo, which asked them to insert small blocks of specific code lines. There is also the concept of a Software Development Kit (SDK) or "devkit", sets of bricks and software tools used in the construction of an application. The highest developments of these tools are known as "software libraries". Installed without your knowledge, a code takes control of your smartphone's system. The firm hidden behind the application can spy on your actions at will, then sell the collected data to contracting actors and others, seeking to make the collection of information profitable. Called "snippets", these codes send the user's personal information to the web, without taking control of the smartphone or the tablet.

5.4.2. *Regulations on data use*

Directive 95–46 was the first regulation on personal data, representing something like oil for the information society. It is not possible to avoid asking the question, because it is still necessary to know how to protect oneself. Among the courses of action to be adopted, we can immediately mention erasing processed data, anonymizing behavioral data or resorting to a pseudonymization of identity, which comes down to separating authentication and identification. A new regulation, called "RGPD", will be in service in 2018.

6 For more information: http://teemo.co/.

In order to guarantee protection, it is possible to use the idea of trusted third parties, as do the actors of Label IFDéNUM. The use of a trusted third party means reducing the number of passwords while handling a strong authentication system.

5.5. The impact of the digital world

The impact of the digital world is most often analyzed according to two thought patterns, which have become classic by now, but which become efficient later, in that the typologies tend to become mixed up.

The first typology is derived from Wasterman (2011). It proposes focusing on three axes, as summarized in Table 5.2.

Axe	Consequence
Temporal	Networks eliminate delays and reduce transportation needs
Spatial	Technologies, especially those of mobility, enable unsuspected ubiquity
Multitudinal	Networks abolish the limits of presence and audience

Table 5.2. *The three axes of digitization*

The second pattern replaces digitization, which not only refers to only some processes but to the whole of productive activities, at the time that it proposes a rather artificial separation. This pattern refers to three transformations.

Transformation	Consequences
Automation	It brings performance forward, which is the ability to individualize the production of goods and services, to increase productivity and to favor a better user experience (see below)
Dematerialization	This refers to the appearance of relationship channels which are different from traditional physical channels. The main consequence is the decrease in marginal costs of production and transaction costs
Intermediation	Intermediation modes are modified, and this modifies not only the value chain (see below) but organizations as well

Table 5.3. *The three axes of digitization*

These patterns are still widely used by university researchers.

In fact, a more promising offer has been developed by the industries: it encourages the creation of a digital double of the factory in order to optimize the production process. The construction of this digital double employs tools that will be described later.

5.6. The digital shift of organizations

Many cases are possible.

5.6.1. *Organizations where the digital shift has been a failure*

Some big firms have missed their digital turn. This was particularly the case of Kodak, which we quoted in the preamble. Some organizations, after their failure, buy other firms. This is particularly the case for banks that buy new players in finance.

5.6.2. *Organizations that made the digital shift early*

Paradoxically, some firms made the digital shift early. In 1977, Nintendo offered a game console which ran on a color television, despite the fact that color had just arrived, and only one household out of 10 had a TV set in the early 1970s. The group, created in 1887, was directed by Hiroshi Yamauchi from 1949 to 2002, the great-grandson of the creator. This man, who was said to be disrespectful of Japanese traditions, was characterized by his indiscipline. It seems that this entrepreneur owed his success to the fact that he began his career in the firm by dismissing the executives appointed by his father. He continued with two master shots: the passage from cardboard to plasticized card and the first authorized license to publish Disney figurines. The idea of the game console started in 1975 after the success of a gadget, a hand that threw marbles. Atari and its game computer were made possible after 1972, thanks to the "Pong" arcade terminal. He decided to position his game console on the model of frugal innovation. ColorTVGame6 was successfully launched in June 1977. Portability was the additional element of the Gameboy in 1989, a pocket console that would sell 118 million copies. His successor launched the Wii console, incorporating a motion detector in 2006. Atari and Nintendo later had a very different destiny.

5.6.3. *Organizations blocked at ICT 1.0*

Among the organizations blocked at the previous level, supermarkets that computerized their payment systems in the 1980s are lagging behind. One of the possible ways is the buyout of a business. This is the case, for example, of the Galeries Lafayette Group in France. It wishes to be inspired by the (spectacular) transformation of La Redoute and has announced the purchase of 51% of the firm's capital. The idea is to accelerate their digital transformation and to transform these stores, by creating a leading omnichannel business group. La Redoute had been left moribund by Kering before its acquisition in 2014 by its executives, for the symbolic sum of one Euro. Nathalie Balla and Eric Courteille transformed the old mail-order society into an e-business.

This firm employed all modern technologies:

– robotization, so as to revolutionize its logistics, reduce order preparation and delivery times;

– m-commerce, so as to be up to date;

– automation, so as to secure sales and ensure better management, particularly regarding logistics;

– virtual reality, so as to seduce the customer;

– Big Data for a personalized customer experience.

Besides, La Redoute revolutionized the way in which it creates its collections in the textile industry. Instead of two or four collections a year, La Redoute moved on to develop 10 collections. It also expanded into other sales areas, such as home decoration. It moved from the edge of bankruptcy to becoming one of the top 10 French e-traders, with nine million monthly visitors and a 2016 turnover of 750 million Euros.

Among the firms blocked at ICT 1.0, let us mention the actors of railway and air transport, whose activity is carried out, thanks to electronic systems whose foundation is still electronic reservation relying on relational databases.

5.7. ICT infrastructure

Internet infrastructures can be appraised according to

– different scales related to the network's geography;

– different service levels, involving a use aspect;

– the protocols used and the applications implemented, which are the technological aspects.

A functional, qualitative and quantitative approach to these infrastructures can be implemented in order to classify these networks.

5.8. Primitive technologies

Primitive technologies remain at the heart of society 5.0. One of them is text analysis.

5.8.1. *Text analysis*

We should bear in mind that the first machines were used for decoding the messages of enemy nations. From this follows that automated text analysis was one of the first applications conceived to be used on calculating machines. This application preceded the development of the first computing machines and those that later became management tools.

After text decoding, two areas were particularly targeted: document analysis and, of course, machine translation. Finally, the task proved to be much more difficult than what had initially been foreseen.

Very quickly, machines were able to decrypt information, which led "the military" to postpone the introduction of other security elements, which have currently become an "encumbrance" for society 5.0.

Translation systems came up against the infinite complexity of languages (ambiguity of contents, impact of paraphrases, polysemic phenomena, etc.). Nowadays, translation systems are of two kinds:

– the traditional approach that employs dictionaries and translation rules. This approach was used in outstanding software, such as SYSTRAN;

– the approach inspired in search engines, implementing statistical models. They seek to identify patterns within large bodies of texts in a context translation (for example, Google Translate employs this second type of approach).

The expansion of the Web, itself linked to an exponential increase in the possibilities of data storage and the processing power of computers, makes it possible to review methods and technologies, of which machine translation is just an example. Thus, the development of the Web has completely changed the way in which problems are managed, and it is this change of solution for solving a problem that illustrates one of the *shifts* related to society 5.0.

For researchers, it is also another way to analyze language. The notions of signified and meaning become obsolete with the advent of sociocognitive dynamic analysis, taking the temporal dynamics into account. The use of Web writings such as blogs, analyzed over a period of time, makes it possible to understand the discussions and to group the actors in types or communities. The major fact is that the observable contents not only become providers of data but also of knowledge, since inferences can be made from the links between these data. Moreover, what characterizes the machine of society 5.0 is that it handles data, at the time that it detects knowledge. Knowledge present in the cyber sphere must be confronted in reality, due to the fact that although machines are effective for obtaining correlations, they cannot certify that these are causalities.

5.8.2. *Voice recognition*

Voice recognition is one of the essential applications of current smartphones, and it is also a technology widely used in trades such as logistics or health.

5.8.3. *The mobile phone as an inclusive technology*

This already-ancient technology had an unsuspected impact on society 5.0, in the sense that it included old technological approaches in its use.

The collective book by Laurence Allard is full of examples of this interlocking of technologies into each other. The mobile phone is a meeting place, not only of people but also something that none of the authors of this book dares to call “a meeting of technologies”. Roger Odin offers an excellent analysis of the meeting between the mobile phone and the cinema[7]. Without it, “an entirety of films would not have been possible”. He rightly observed that the mobile phone did little to change the cinematic technique, but it significantly influenced narration. It enabled the interactivity with the viewer, which added a component to the photographic display. Benoît Labourdette insisted on a major difference with other technologies concerns the audience, which opens up and widens. If watching a Super 8 film was previously limited to the family field, the characteristic of the mobile phone is that, since around 2005, it has broadcast productions worldwide, via platforms[8]. In a previous chapter, we highlighted the importance of platforms for society 5.0. The modification of storytelling was conceptualized by William Uricchio: “Digitalization, networking and accessibility, these three factors allowed intense participation, expanded the notion of data to include localization and fundamentally changed the notion of documentary topic”[9].

Thanks to the applications introduced in this multifunctional mobile phone, it has now become common to talk about the power of the “button”. The example of YouTube is quite enlightening. YouTube has become a video fountain, with more than a billion hours of content per day, 400 hours of content loaded every minute. As I am writing this text, YouTube users are uploading over 100 hours of online video every minute, and Facebook is pushing more than a billion users to upload their daily multimedia autobiography to their personal page (photos, video clips, geolocation maps, links and diary pages).

5.9. Recent technologies

Different technologies have become crucial in society 5.0.

5.9.1. *Robotics and automation*

Robotics is also a feature of society 5.0. Robotics is not the preserve of developed countries. An Indian start-up created in 2011, the firm Gray

7 Laurence Allard, Laurent Creton, Roger Odin (dir.), *Téléphone mobile et création*, p. 38, Armand Colin, Paris, 2014.

8 Laurence Allard, Laurent Creton, Roger Odin (dir.), *op. cit.*, p. 55–59.

9 Laurence Allard, Laurent Creton, Roger Odin (dir.), *op. cit.*, p. 62.

Orange,[10] flooded the world with its logistic workshop robots. The first one, Butler, can intelligently store more than 600 different goods per hour and the second one, Sorter, can assemble, sort out and scan packages with unparalleled speed.

5.9.2. *Virtual reality*

Virtual reality can be used for training. Different reasons are given for this: training for complex tasks, access to very expensive or hard-to-use equipment, preparation for dangerous environments and employees who are geographically scattered. Kentucky Fried Chicken (KFC), the fast food brand, announced a food preparation video. UPS recently announced the integration of a virtual reality module into its training centers. Walmart intends to train 150,000 virtual reality employees every year. Many of these trainings can function as vectors for other actions:

– from a marketing viewpoint, it can give customers an image of seriousness and professionalism;

– it can publicly praise the creators of the firm. In the example of KFC, the role of insane torturer is played by Colonel Sanders, the founder and mascot of the brand;

– a form of video game for seducing employees.

5.9.3. *Computer-aided design*

Computer-aided design (CAD) software is present on the market with different applications and features.

Design software is great for building complex surfaces, but it is still a bit rigid. These programs include 3Design, Alias, Cobalt and Evolve. In France, the reference is Rhinoceros 3D and is used by many design studios. This software makes it possible to produce very complex surfaces with maximum flexibility. Let us observe that 3Design was originally dedicated to jewelry design.

10 For more information: http://www.greyorange.com/.

Direct modeling CAD software evolved from 2D to 3D. AutoCAD, a now historical software, is the most widely used for designing 2D plans. AutoCAD has evolved with 3D features. It is broadly used in the architecture trade. Let us also mention BRL-CAD, CorelCAD, CREO Direct, Draftsight, eMachineShop, IronCAD, SpaceClaim and TurboCAD.

The so-called "parametric" CAD suites are the oldest ones. These suites are used by most industrial designers and mechanical engineers, as well as some industrial designers. They are the historical software of this type of application. While Solidworks has the merit of having a very intuitive interface and being the best known, ALibreCREO (ex ProEngineer), FreeCAD, Fusion 360, Inventor, NX (Unigraphic), Onshape, SolidEdge and TopSolid do not have the same advantages. Their use is more complex, but they are more complete, which means that they are intended for professionals whose digital transition was made possible, thanks to the respect of their rules and codes. Finally, Catia is the reference for the big industry. Catia has extremely advanced options and is accompanied by many additional modules that are suitable for all industries. One of them makes it possible to design factories.

Simplified 3D modelers also exist. They produce 3D shapes without focusing on the plans and are limited in their functionality. Let us mention the famous Sketchup, as well as 3D Slash, Anim8or, Amapi, Art of Illusion, K-3D, Metasequoia Moment of Inspiration, Silo, ThinkerCAD and Wings 3D.

3D infographics and animation software do not produce plans, but synthesize images, either fixed or animated ones. The forms produced are usually photorealistic and can often be animated. Blender is the reference in Open Source. Houdini is a very advanced special effects software which focuses on the procedural generation of 3D objects. 3D Crafter, 3D Max Studio, Cheetah3D, Cinema 4D, 3D Lightwave, Maya, Modo, Mudbox, Sculptris, 3D Shade, Strata 3D Design and ZBrush are among the software of this kind.

In order to create and animate people or animals, Daz Studio, MakeHuman, Massive, Mixamo and Poser can meet the demand. Bryce, Terragen or Vue will make it possible to easily create landscapes and natural scenes. 3DVIA, or SweetHome3D, helps you visualize interiors and optimize layouts, shops and boutiques in particular.

5.9.4. *Artificial intelligence*

Artificial intelligence (AI) should make it possible to

– improve customer satisfaction;

– propose co-constructions of innovative services or products;

– increase productivity gains in organizations.

AI is typically one of the tools of society 5.0. Three types of AI can be distinguished.

Computer programming based on algorithms is the first type. An algorithm is a mathematical model translated into a sequence of instructions and processes which are necessary for performing the action. Among this type of algorithm, the most frequent ones make it possible to predict the evolution of an indicator from a palette of numerous variables. Personal wizards from Google (Google Now) and Apple (Siri) are integrated with their software proposals. These are also present as chatbots included in the applications.

There is a second type that relies on much more complex algorithms, using rules and databases. These are called “expert systems”. By means of a special process, the creators of this software included knowledge on regulations that was subsequently applied to data. These programs won against chess champions with error effects which were generated or forgotten in these programs. Evolution has taken place, thanks to the introduction of an automatic learning mode. The advantage of expert systems is that they make simulations possible and are used in autonomous vehicles (cars, trains, planes, etc.). The recommendation engine developed by the Amazon selling tool is based on the data of its users and falls within this framework.

Evolution refers to software which can learn on its own and correct itself. Machine learning is the name given to these programs. Google’s AlphaGo, which defeated Go’s World Champion, or IBM’s Watson, which won the game show Jeopardy, are examples of this. Currently, very real and useful applications are poorly developed. Human facial recognition, or object recognition, and the understanding of natural language are favorite areas.

Start-ups or large firms like IBM and Microsoft offer such software bricks. Smart Me Up has developed a facial recognition solution. It offers the possibility of monitoring passenger flows in the station or detecting a driver's somnolence. Moodstocks is a smartphone application for identifying objects, thanks to its integrated camera.

6

Society 5.0 and the Management of the Future

One of the major elements of future management is the importance of data. This is going to determine new forms of marketing and the establishment of "servitization", which we will explore later.

"Servitization" relies on a new marketing base, in which needs replace the analysis of demand.

6.1. The firm from the managerial viewpoint

While management refers to "the firm's vision", marketing employs a particular definition of the firm.

The latter has a "societal" and an economic foundation which can be simplified by answering two questions: why and what for?

6.1.1. *The definition of management*

DEFINITION. Management is a process which involves making decisions, setting goals, organizing resources, directing them and monitoring the achievement of goals. It corresponds to the concept of planning, as conceived by Fayol.

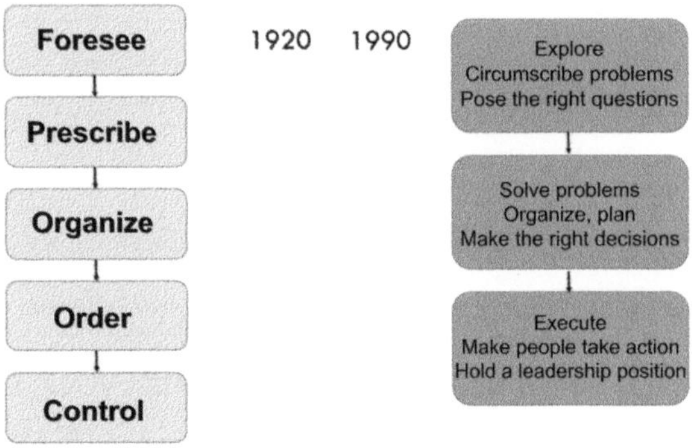

Figure 6.1. *From Fayol's management model to the modern model. For a color version of the figure, see www.iste.co.uk/salgues/society.zip*

6.1.2. *Management's contents*

Management is based upon different elements which are summarized in Figure 6.2.

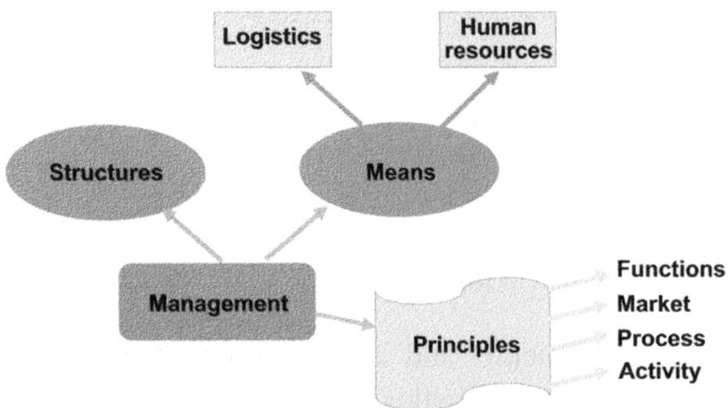

Figure 6.2. *Management components. For a color version of the figure, see www.iste.co.uk/salgues/society.zip*

6.2. The definition of market

DEFINITION. The market is a meeting point between supply and demand. As such, supply and demand evolve hand in hand with the concept of management, etc. itself stemming from changes in the environment.

In Western Europe, the majority of authors identify different phases in its evolution.

Business type	Period	Marketing features
Taylorist firm	Previous to 1945	Demand > supply Production increase
(Fourastié: The Glorious Thirty) Market research Segmentation	1945–1975	The Glorious 30 Marketing
Customer's will Cost and economic conditions	1975–1990	The 15 Dark Finance and management control
Supply > Demand	1990–2000	The authentic years Personal management
Possible Supply > Demand and possibility of obtaining information about a potential customer	Since 2005	Customer Service Datum (Big Data)

Table 6.1. *Different phases of management*

The questioning of marketing is a necessity in society 5.0. It is characterized simultaneously by new forms of customer knowledge and a necessary abandonment of the traditional forms of marketing.

6.3. Marketing

Traditional marketing will undergo certain transformations in order to remain useful in society 5.0.

6.3.1. *Marketing is an approach which only makes sense in a certain context*

Marketing was developed within the industrial society, particularly during the crisis of the 1930s. It is a management tool at the service of firms and organizations, placed in the context of a market economy, seeking to increase performance. This situation corresponds only very partially to society 5.0. Marketing methods are changing as a consequence of changes in new firms.

In the history of marketing and its evolution, several definitions have been significant: those of Converse, Joannis and Kotler.

6.3.1.1. *Marketing according to Converse (1936)*

DEFINITION. "Any service, act or operation performed at the production or distribution level, in order to satisfy the desires of the consumer".

This definition of marketing focuses on consumer knowledge. It refers to knowledge, not to understanding.

6.3.1.2. *Marketing according to Joannis (1965)*

DEFINITION. "Marketing is a "scientific" method for detecting and profitably conquering the market".

This definition focuses on detection by means of studies and the concept of increased performance. Economic balance is one of the new factors introduced. This type of marketing presupposes the existence of market studies, which may no longer be meaningful in society 5.0. Data directly taken from the customer at the time of purchase are more important than those resulting from studies and therefore prevail.

6.3.1.3. *Marketing according to Kotler (1986)*

DEFINITION. "Marketing management is the analysis, planning, implementation and control of programs designed to create, develop and maintain a mutually satisfactory flow of exchange with target markets, in order to achieve the goals of an organization. It is essentially based on the creation of supply in terms of needs and on the development of a price, communication and distribution able to inform, motivate and serve the market".

This last definition introduced management and marketing mix. Marketing mix is a technique for defining the product, its price, its distribution method (placement) and its associated communication modes (advertising). Marketing mix is also known as 4P. Technical know-how and commercial action are not enough to ensure the success or the mere survival of the firm when confronted with the economic challenge of current markets.

6.3.1.4. *Marketing tools*

In an industrial society, it is necessary to take not only the market's influential forces into account but also the firm's strengths and weaknesses, as well as the market's threats and opportunities in order to

– integrate foreseeable developments;

– adjust supply to demand, or vice versa[1];

– defeat competition, or, at least, obtain more favorable survival conditions.

Let us recall here our initial assertion introducing the idea that marketing is an approach which only makes sense in a specific context: that of the existence of a firm and a market.

As with management, different marketing periods can be identified chronologically.

6.3.2. *The four historical periods of marketing*

In order to analyze marketing actions, it is useful to place society in relation to the four phases of marketing's maturity, or "phases". These phases emerged concomitantly with four historical evolutions. Nowadays, it is possible to find firms that respond to each of these types.

6.3.2.1. *Phase 1*

The firm only intended to sell its production on the market. The establishment of distribution and sales channels was the only important element.

6.3.2.2. *Phase 2*

Flow problems emerged, and the firm became interested in this problem. In this way, marketing thought turned toward distribution.

1 In many industries, marketing is expected to shape customer demand in order to match the optimum production of factories. This idea was later transmitted to the services industry.

6.3.2.3. *Phase 3*

The set of marketing variables became important since a firm had to place itself in the market. Supply was greater than demand, and firms were forced to outperform their competitors.

6.3.2.4. *Phase 4*

Not only did firms become involved in exacerbated competition, but they had to learn how to anticipate it. Therefore, they became necessarily obliged to get to know their customers and also to understand them.

6.3.2.5. *Phase 5*

Business operations are becoming increasingly based on customer data, and firms start working as data providers.

Society 5.0 represents a challenge to the improvement expected from each phase and aims at perfecting each phase as needed.

6.3.3. *The features of the different phases*

6.3.3.1. *The flow phase or phase 1*

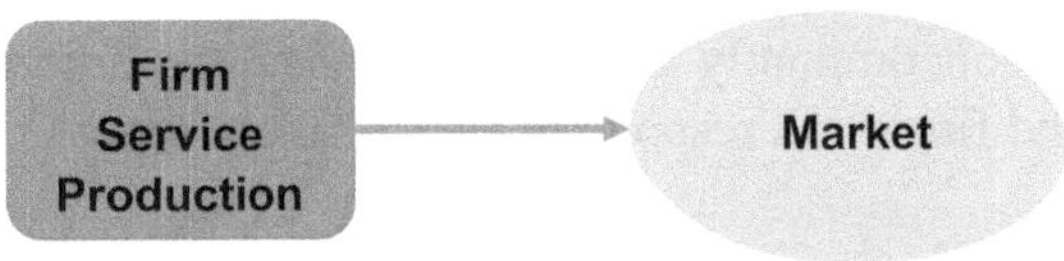

Figure 6.3. *The flow phase. For a color version of the figure, see www.iste.co.uk/salgues/society.zip*

The market, as compared to production, is a simple way of selling the products manufactured by the firm. In fact, no attention is paid to the marketplace. Demand is subject to the situational developments of the market. In general, demand is greater than supply: there is market penetration. Nevertheless, sometimes, some products have a demand which is almost equivalent to the supply. In this case, the situation is delicate. The revival of the idea of short circuits, especially in the field of agribusiness, is one of the examples where society 5.0 uses phase 1 to its advantage.

Let us quote Ford's famous phrase:

> "Our policy is to reduce prices, develop sales and improve our product. We never consider fixed costs as such, so we start by reducing the selling price to the level that we think it is suitable for stimulating customers... we do not break our heads over production costs, the new price will force them to go down... since how could it be useful to know the costs if the fact of knowing them teaches you that you cannot produce at the price in which the product will be able to sell?"

In this case, the price of products is determined by the production costs and the marketing action is not executed by the prices. Society 5.0 is placed in this context, especially in services where staff costs represent an important element of production.

6.3.3.2. *Product commercialization or phase 2*

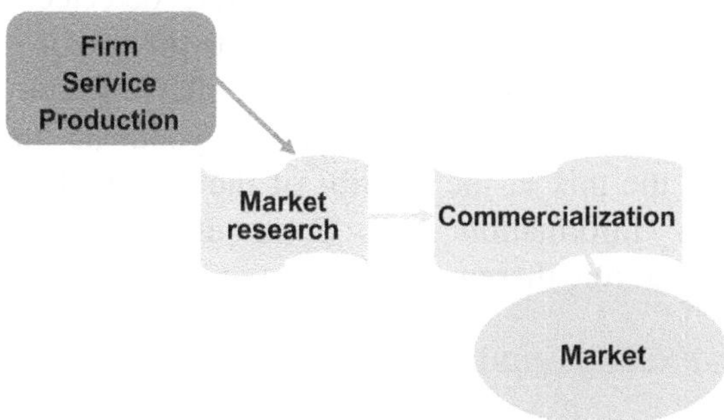

Figure 6.4. *Marketing phase 2. For a color version of the figure, see www.iste.co.uk/salgues/society.zip*

Market research (see questions on the studies) is carried out after the launch of a product; its only purpose is to seek outlets for the products manufactured by the firm. Commercialization is characterized by a certain commercial dynamism which reveals a will to sell an entire production, not only to sell scattered products, as in the first phase.

6.3.3.3. *The coordination phase or phase 3*

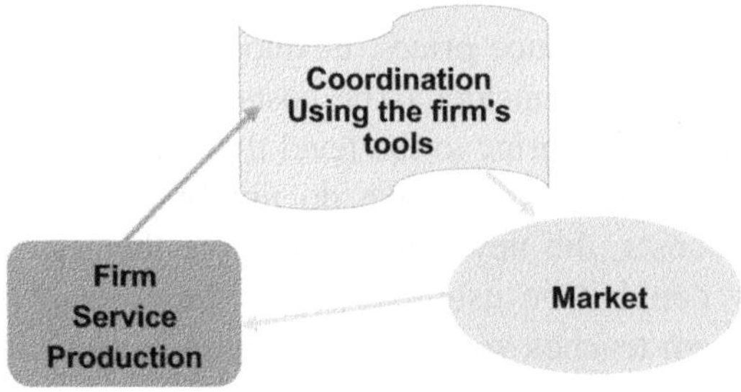

Figure 6.5. *Coordination phase. For a color version of the figure, see www.iste.co.uk/salgues/society.zip*

Phase 3 was dominant during the *Trente Glorieuses*[2].

The firm's production is strongly conditioned by taking into consideration the needs of the market that were detected through market research. It is also the market which later orients the use of marketing tools: marketing mix (price, sales force, distribution, after-sales promotion). In the context of society 5.0, coordination extends beyond the boundaries of the firm, determining the mix at the interior of the firm with its computerized production tools. Information and Communication Technologies 2.0 (ICT 2.0) play a vital role therein.

6.3.3.4. *The marketing approach phase or phase 4*

Marketing becomes both a logic and an approach. This action has its origin outside the firm, thanks to the understanding and knowledge of the market. It originates in the firm's design of its commercial offers and its actions at the market. It is characteristic for its validation and overall consistency.

2 A period in France's economic history following the Second World War (1945–1975).

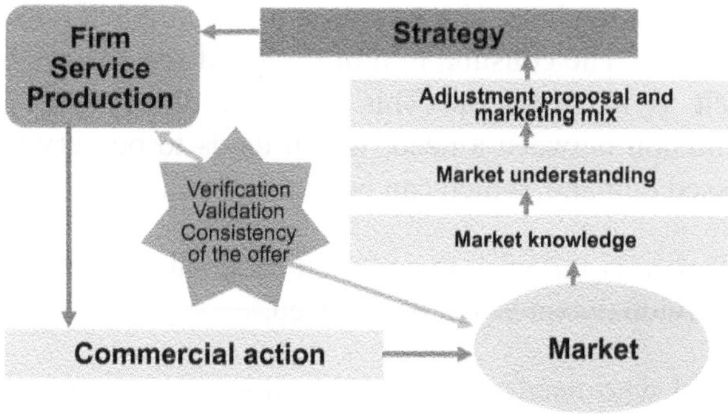

Figure 6.6. *The marketing approach phase. For a color version of the figure, see www.iste.co.uk/salgues/society.zip*

At this phase, understanding markets becomes as important as their knowledge. The speed of changes is a feature of this phase. If there is an error on the phase (coordination or use of the firm's tools), there is a swift disappearance of the market between one and three years. Numerous examples exist: the furniture sector in France witnessed the disappearance of its leader, the Barbès Galleries, whereas fast food hamburger chains have eliminated small restaurants, etc.

Society 5.0 takes full advantage of the use of this phase in the cases where consumer feedback is easy and thus feeds both market knowledge and its understanding. The difficulty comes from the implementation of these tools.

6.4. The logics: need, desire, expectation and demand

Need, desire, expectation and demand are the four concepts that are closely intertwined in the marketing approach.

6.4.1. *The Lacanian perspective applied to marketing*

The Lacanian vision has defined a chain: need, desire, expectation and demand.

DEFINITION. A need is a lack, felt as dissatisfaction.

Eating, drinking, sleeping and communicating are the basic needs, necessary for man. Actually, a need points to a real object, but envisions it by means of a decision chain. Consumption of the targeted good is measured in terms of demand. The consumption of the good results in satisfaction, the opposite is felt as a frustration. This direct relation to the good quickly entailed a confusion between a need, which tends to be very stable in man, and the expressed demand, which can be highly variable.

DEFINITION. Desire is the wish to see such need satisfied in a specific way. Desire is an individual expression of the need.

An individual who needs to eat might rather desire a *pot-au-feu* or raw fish, depending on whether he is French or Japanese. Desire is linked to a lost experience, a lack that cannot be filled.

DEFINITION. The term "expectation" refers to what the consumer "expects" from the product. It is the expression of the global vision of the product, not just its technical or functional vision.

When a man buys a powerful and fast car, this can be an answer to different expectations: it can be an expression of social status, it can be chosen because it goes fast, because it "seduces" the girls or because it arouses the jealousy of his colleagues. The notion of expectation should be understood in relation to the type of society we live in and therefore represents one of the key concepts of society 5.0.

DEFINITION. We speak of "demand" in the case of a desire for a specific product associated with expectations, with the possibility and the will to acquire it.

Demand is the only directly observable variable; it is also the key variable of economy. The term "possibility" refers, for instance, to purchasing power, presence at the point of sale, compatibility with preexisting equipment, etc.

6.4.2. *The place of marketing*

Auguste Walras made a distinction between morality and political economy. For him, morality describes as "useful" only those objects that will satisfy the needs declared by reason, whereas economy refers to all the objects that man may desire, either in the interest of his preservation or by the effect of one of his caprices, or one of his passions.

6.4.2.1. *Need*

Contrary to popular opinion, marketing does not create needs. Some proverbs result from the analysis of the logic need, desire, expectation and demand. There are no products as such, expressed by a sole demand, there are only consumer needs. There are no consumers as such, nor even users, there are only buying processes. Identifying the consumer or the user is to position him in relation to the different phases of the purchase process. There are no prices as such, there are only distributive systems. Thomas Aquinas stated that man seeks goods only in order to satisfy his needs. He made a warning against vice, the immoderate love for possession.

6.4.2.2. *Desires*

Every human being has desires. When these are not satisfied, man becomes sad, he seems confined in his body and soul, he feels oppressed[3]. Certainly, desire is never the result of knowledge, of awareness, but of an impulse. In other words, desire is not rational, and we experience desire before knowing what we desire. Spinoza made desire the essence of man and perseverance the essence of desire. It is for this reason that desire takes place in time. Liaudet recalls that while the myth is like a dream, the disguised expression of desire, it is still necessary for it to go hand in hand with reality[4].

Schopenhauer considered that life oscillates between pain and boredom. When we suffer from lack, and we desire things, we feel unfortunate. When we fulfill a lack, we feel unhappy because we no longer desire.

6.4.2.3. *Satisfaction*

Consumer behavior specialists seek to satisfy the consumer by building loyalty to the product, service or brand. The existence of a usage value assumes that buyers and consumers are able to identify the use of a product or a service and value it for what it is. In this approach, this involves the identification of functions. All the authors insist on the fact that the identification of this function is not enough to guarantee adequate, efficient[5] consumption. The method is simple: we identify the functions that make it possible to recognize the use and appropriation makes it possible to define

3 Bertrand Vergely, *op. cit.*, p. 9.

4 Jean-Claude Liaudet, *Le Complexe d'Ubu ou la névrose libérale*, p. 31, Fayard, Paris, 2004.

5 Ladwein, *op. cit.*, 2003, p. 359.

values of use which translate customer satisfaction. This traditional view is too restrictive. As we see later, it does not define all the uses of the products.

6.5. New managerial skills

New managerial skills will be needed in society 5.0. These are introduced in Table 6.2.

Skill	Description
Innovation economy	Understanding the new forms of innovation (including social innovation and frugal innovation)
Digital writing	Becoming trained in writing practices for digital media and to the options that these make possible
Image	Understanding the importance of the image that the organization conveys, using and implementing tools and people (ambassadors of image) for distributing it
Artificial intelligence	Understanding artificial intelligence tools and their limitations, as well as knowing how to use them, particularly in cases of access to voluminous data (Big Data)
Collective intelligence	Knowing how to use collective intelligence and setting up events or tools so as to capture it, carrying out actions capable of grasping cultural differences
Lobbying and influence	Ability to implement actions
Creating attention	Being able to catch the other's attention
"Co" planning	Implementing plans for developing forms of collaboration, co-construction, co-design, etc. The idea is to favor "Co" society
Globalization terminology	Understanding the terminology of globalization, particularly that of logistics (INCOTERM)
Automated work and robotics	Understanding which part of the activity may be subcontracted to machines, understanding its advantages, strengths, weaknesses, opportunities and limits, understanding the role of data capture for automation
Competitive intelligence, economic intelligence and mash-up (see below)	Implementing tools for collecting and analyzing data Implementing aggregation tools

Table 6.2. *The skills needed in society 5.0*

The Jugaad experience accompanies entrepreneurs and managers who want to evolve toward a liberated society, by means of management based upon trust. It also helps firms to develop internationally and boost their creativity, thanks to Jugaad Innovation or intercultural intelligence.

According to Navi Radjou, frugal innovation refers to achieving more with less. An innovation consultant in Silicon Valley, he is the father of Jugaad. As a movement, Jugaad was named after a word in Hindi meaning "knowing how to manage and find solutions in the middle of hostile conditions". It is a real state of mind in the sense of a marketing mindset[6].

6.6. Boredom comes from repetition

"A great deal of boredom is due to repetition"[7]. Our reference, the consultant Nathalie Joulin, suggests two essential remarks. First, we should know how to manage small innovations. Then, invention, and often reinvention,[8] can take place. Here, we can appreciate the role played by boredom and the difficulty that this consultant encountered in the indefinite use of the term "innovation": she thus felt compelled to substitute it for "invention".

6.7. Customer satisfaction

The firms in society 5.0 will have to organize themselves in order to better satisfy customers, by using relevant data. Society 5.0 does not represent a resistance to consumption, but the will to improve consumption. Besides this, we are confronting an environmental challenge that cannot be avoided. The relationship between supply and demand changes. Society 5.0 is characterized by the fact that actors are abandoning the transactional market archetype, associated with commodity exchange against money and moving toward a relational market, where "servitization" and the use of "Co" are key elements (these themes will be addressed later).

---

6 For more information: Simone Ahuja, Jaideep Prabhu, Navi Radjou, *L'Innovation Jugaad*, preface by Carlos Ghosn, translation by Jean-Joseph Boillot, Dataneo, Paris, 2016.

7 Lars Fr. H. Svendsen, *op. cit.*, p. 56.

8 Nathalie Joulin, *op. cit.*, p. 61 and following pages.

Package, solution, experience and useful effects, etc., the product is merely the support. These new markets are moving the focus out from consumption, while the exchange value is migrating toward use value. Transactions are often referred to as "transactions in exchange for money". In fact, consumers exchange a good or a service against currency. When we switch to the value of use, the transaction itself changes. The business relationship takes place on both sides of the transaction, the diagnosis occurs upstream, where information is exchanged at first, and then flows downstream, subsequently verifying the useful effects of the acquisition of the property or service. Thanks to the advances of the digital world, the density of the relationship is increasing. The customer is invited to provide information, consciously or not (the Internet of things occurs without his knowledge), so as to exploit the situation of economic dependence in which he is suddenly placed.

The emphasis on the business relationship requires a redesign of relational architectures, including the concept of platform, which will be discussed later. A legal wall emerges, Law must admit that the customer is held captive and should sanction abuses.

6.8. Resistance to consumption

The marketing of society 5.0 is confronted against consumer resistance. This society considers itself performative and has little interest in resistance as well as to what consumers can do against this. Resistance results from marketing discourse and the differences between speech and behavior, which is increasingly becoming a requirement for recognition.

The consumer's resistance takes the form of audible, collective and visible movements, such as boycotts, smear campaigns, consumer happenings against mass distribution, or less audible demonstrations, such as identity manipulation, collective complaints, the will to boycott consumption and the voluntary search of simplicity.

At the beginning, the attacks were targeted against firms. These were prompted as calls for boycott. Later on, they evolved toward the contestation of established principles. This meant an escape from traditional markets that increasingly shifted toward community-based approaches or were aimed at respecting the environment. As an example, we can mention the opening of shops only selling recycled products in the European countries.

6.9. Recovery, gleaning, etc.

Information retrieval and the gleaning of informational items are new actions related to the actors of society 5.0.

In the face of these actions, different forms of opposition have emerged, and these correspond to situations where pressure may be real or perceived. Man is considered to be free, except that he is "oppressed", meaning that he is autonomous while he is permanently helped, and it is precisely this aid which leads to the situation of oppression. Promise must be weighed against the reality of supply, particularly in what concerns digital tools. The major risk is that the customer may perceive this as a betrayal (Ward and Ostrom, 2006).

Current information systems use an overflow of identifications, which suggests a lack of consideration, an absence of gratitude. This leads to the emergence of coalitions, media coverage phenomena and punitive movements. The risk is that business discourse and other related manifestations may become united against business.

6.10. Customer relationship management: an essential tool

Customer Relationship Manager (CRM) is a necessary computing tool, regardless of the activity sector concerned. CRM is a tool for managing business organizations. It helps to improve turnover and save time to employees. Thus, this tool can help to improve profitability.

CRM centralizes information concerning a customer and keeps track of trade relations, as well as the progress of ongoing projects. The tool makes it possible to list actions that have already been carried out or still have to be performed. It makes it possible to assign tasks and to validate decisions. Generally, it can help identify obstacles. CRM also makes it possible to generate dashboards according to different views: the customer's view, the pipeline view, the project view or the performance view, which measures progress and performance ratios. Society 5.0 is increasingly requiring better customer service, for which the customer relationship tool is an unavoidable tool. Tracking modules help to measure the performance of commercial actions and to incorporate lead scoring in such a way that marketing efforts are reflected on those customers to whom the organization is most likely to sell. All these actions contribute to define relevant indicators. CRM also promotes teamwork with common agendas, collaborative documents and a history of interaction.

6.11. The holistic approach to management

Society 5.0 features a holistic vision of management. This involves abandoning structures reflected beforehand as well as the sources of cooperation costs and expenses related to necessary connections and consistency. This is why the actors of society 5.0 have suggested building a corporate structure based on interpersonal relations, an approach that received the name of "sociocracy" or "holacracy".

6.11.1. *Sociocracy*

Sociocracy is the oldest model of management 5.0. It dates back to the work of systemics, in vogue during the 1970s.

DEFINITION. Sociocracy is a mode of governance, based on simple tools, which allows an organization to operate effectively without a centralized power structure, according to a self-organizing mode which disseminates decision making.

What differentiates sociocratic organizations from each other is the implementation of tools for setting up organizations. Its main tools are

– consent governs the decision-making process;

– the election of decision makers with no candidates;

– the organization of interconnected autonomous functional units;

– double linking of individuals between circles.

The foundation of sociocracy is the creation of functional units, which the founders designated by the term "circle".

The term "sociocracy" was invented by Auguste Comte. The construction of the word combines Latin expressions such as *societas* (society) and Greek ones, such as *krátos* (authority). Auguste Comte wanted to promote the governance of the *socios*, that is to say, of people bound by meaningful relationships. For example, Auguste Comte wanted to separate this approach from the one stemming from *dêmos, a* mass of individuals who share common values, in other words, democracy. This author did not envision this as an idea of organization, but as a new political form. "To free the western hemisphere from an anarchistic democracy and a retrograde aristocracy, to constitute a true sociocracy, as far as possible, wisely contributing to the common regeneration of all human forces"[9].

In 1926, Kees Boeke,[10] founded the Werkplaats Kinderge-meenschap (WP) school in Bilthoven, the Netherlands. He was one of the first to use the most innovative teaching practices at the time, inspired by the theories of Maria Montessori. The rules of this school are based on decision making based on consensus. Gerard Endenburg, a theorist of sociocracy, was a student of this school from 1943 to 1948, particularly during the war period.

In business settings, this approach has been popularized by consultants. A French example is the IDEOdynamic Institute[11].

6.11.2. *Holacracy*

Holacracy might be interpreted as an evolution of sociocracy, but actually, it is that implementation that differentiates holacracy from the other method.

DEFINITION. Holacracy is a governance organization system, based on a formal use of collective intelligence. From an operational viewpoint, it is a method that makes it possible to disseminate decision-making mechanisms throughout an organization by forming interconnected and self-organizing adequately sized groups.

---

9 The theory is defined in Auguste Comte's book, *Catéchisme positiviste*, p. 2, Garnier-Flammarion, Paris, 1966.

10 Kees Boeke, whose real name was Cornelis Boeke, was born on September 25, 1884, at Alkmaar (Netherlands) and died on July 3, 1966, at Abcoude.

11 The IDEOdynamic Institute was created by Fabien and Patricia Chabreuil, who teach sociocracy and who have used it for consultancy and coaching since 2009. For more information: http://www.sociocratie.net/.

6.12. The hacker's position

The most successful organizations seek to use the hacker's methods and tools, that is to say, hacking.

6.12.1. *Corporate hacking*

According to Pascale Gentil,[12] nowadays, corporate hacking represents the world of creative disruption. The hacker, who loves finding solutions, was born in the United States in the 1970s, together with a specific vision of the hacker.

Access to computers, and anything that can teach us how the world really works, should be unlimited and total.

Information should circulate freely and be free of charge.

Beware of authority, encourage centralization.

Hackers should be judged according to their works, and not according to fake criteria, such as position, age, nationality or qualifications.

We can create art and beauty on a computer.

Computers are made to change life.

Adapted from a text by the Massachusetts Institute of Technology (MIT), available at: web.mit.edu/ and http://hacks.mit.edu/Hacks/misc/ethics.html.

Box 6.1. *The ethics of the hacker*

With the arrival of these ethics to the firms, the watchword is innovation. Conformism allows for rules, but it represents stability. In a changing world, the term "routine" is no longer of use. "Be the change that you wish to see in the world". This phrase, attributed to Mahatma Gandhi, is the leitmotiv for hackers.

12 Pascale Gentil is the creator of Imfusio, a consulting firm in style change. Having worked at the Human Resources Direction and with a background in Law, she decided to become a hacker, in order to be more efficient.

Let us quote several important points mentioned by the actors involved, in particular those stemming from Margaret Mead's observations, who proposed an important idea on this subject:

> "Never doubt that a small group of thoughtful, committed citizens can change the world; indeed, it is the only thing that ever has".

This sentence means that a small group of thoughtful and committed citizens can indeed change the world. In fact, it has always been that way. Other sources of inspirations are used by firms: the following three approaches are some examples.

6.12.1.1. *Everett Rogers' Innovation Diffusion Curve*

The innovation diffusion curve for new products was introduced by Everett Rogers in his book *Diffusion of Innovations* in 1962. This theory involves two aspects. The first aspect associates different groups of customers with the different adoption phases of a new product. The second aspect seeks to focus on the conditions that favor innovation. The best-known example is the keyboard of our modern computers. Failure to generalize a more efficient keyboard has been emblematic of the significant dependence on the technological path. Past decisions made throughout the history of technology are taking root and are limiting new developments.

6.12.1.2. *Otto Scharmer's Theory U*

U-procedure is based on a concept that Otto Scharmer called "presencing". This is a neologism formed from the words "presence" and "sensing". U-procedure is a metamodel which makes it possible to shift from a State in which we ignore where we are going (that of weak signals and the emerging future) toward the new social States of the world. These are situations in which new products have replaced the old ones, even if these are still present on the market. This approach offers methods for rebuilding the tools we already have, to reorganize them and to carefully watch their performance. Theory U is the resumption of a phenomenon known by economists as the J curve.

6.12.1.3. *The tipping point by Malcolm Gladwell*

The tipping point should be understood as the "toggling point". In other words, it defines the smallest thing that can be done in order to positively hack a system. It is the very precise point that enables shifting or bissociation.

The mourning curve applied to corporate management is represented in Figure 6.7.

Figure 6.7. *The mourning curve and corporate management. For a color version of the figure, see www.iste.co.uk/salgues/society.zip*

The observation is the following: the mourning curve involves constraint; this mode of management is a source of distrust. That is why current organizations are suggesting relying on people who want something.

The consulting firm Imfusion invented the Labo, Proto, Transfo model. The Proto phase precedes the dissemination of practices, dissemination that takes place in zones of influence. It is the tipping point that triggers transformation. For this, it is necessary to abandon the siloed organization model, which adds unnecessary layers. The underlying concept is to build a product, a service or a process with a small number of individuals, around 5% of the actors in the Lab phase, and then to move to 20% of the actors in the Proto phase, which stabilizes and tests the model out. Finally, once the Transfo phase is reached, its effects are propagated throughout the whole firm.

In concrete terms, the method is divided into four points. The action encourages the individual to experiment and to experience rules, instead of striking against them. As well, the promoters of these actions believe in the power of collective intelligence. Thirdly, the concept of inspiring travel is an effective method, since working on the trip leads to an integration of the previous operation mode. Finally, it prompts acting without resistance, by making the need for letting go explicit and suggesting working on the Self.

6.12.2. *Managing a hacking session*

A hackathon is often conceived as a two- to three-day event in which the employees of an organization must find new ideas for creating a new product or a service, or for the internal or external work of their organization. This restricted time frame does not make it possible to draw maximum benefit from the operation, but this work has a healthy start or initiation. As we see later, the ideal is to "hack" the entirety of the organization, for it to be situated in a privileged position at society 5.0. The other difficulty is to make sense or to prove the usefulness in the innovation of such product, service or operation for the way in which the organization works.

For a hackathon to be successful, it is necessary to clearly define the objectives, first. It is necessary to go beyond the idea of a meeting in order to create new projects. However, establishing too narrow a goal is to risk disappointment. The risk is twofold: participants may be disappointed if the projects are not reused by the organization, and the organizer may find that the results do not meet expectations. The goals should be broad enough: taking the Open Innovation of a product as a point of departure, testing new work methodology, collaborating between units, etc. Nevertheless, if sometimes the operation does not meet expectations, it should be remembered that this type of event creates cultural changes in the organization.

Giving a tool box to the teams that will participate is as important as putting people together. The idea is to create a proper environment for working together. Among these tools, we can include "coffee conferences" or "design thinking workshops", in order to raise awareness among the participants and make them sensitive to the future recipients of the innovations. The meeting of resource contacts is also useful. Finally, it is necessary to provide modeling tools, which are generally found in living labs.

When an event is scheduled, it will probably have a higher added value. Creating living lab-type tailored workspaces is not enough. It is necessary to ensure that the actors are cut off from their workplace. They should feel as if they are in another universe. This universe must enhance daring and push workers to conceive original questions. The presence of facilitators will help ideas emerge and redirect reasoning paths.

The last step is to widely acculturate the organization in order to change the mindset of the participating actors, for them to become more innovative and responsive to society 5.0. Methodologies, products, ideas and tools experienced at these events should be reusable in daily life. Digital acculturation is one of the points of a larger transformation plan. Other events may be complementary so as to facilitate change: digital days, learning expeditions, startup days.

It might also be necessary to find ways that could extend the results of these events. Doing nothing with the results obtained is always regrettable. It is sometimes difficult to further them after the return to normal life. That is why it is often necessary to carry out team building activities and to take participants away from their department so as to bring them into new structures which motivate entrepreneurship within the organization, which enhance competitiveness and invite mastering new tools.

6.12.3. *Human resources management*

One of the first messages is the following: the first person to change so that the business changes is you.

The three axes of human management for organizations are described in Table 6.3.

Axis	Observations, goal
To place the human at the heart of organizations	A firm is not merely capital which has to be set in motion, but it also comprises people who need to be managed
To make work meaningful	This involves generating autonomy, creating new forms of engagement, ensuring the acceptance of visions and objectives
To attain effectiveness	This involves doing the best for a sustainable scenario to be attained

Table 6.3. *Human resources management*

Consultants propose tribes without a hierarchy, but sharing discussions and decisions, as well as transparency. It is necessary to avoid competition among the actors of the organization.

For their own society, they set up a conscientious remuneration, with a choice of salary, including associates. This generates self-confidence and some pleasure. The process goes through various phases: a salary vision inspired by benchmarking, a reflection about money, interviewing two people to generate a mirror effect, adjusting the desired remuneration, a public announcement of what is wanted, asking questions to understand and clarify choices, overall adjustment, veto potential that might jeopardize the balance of the collective if it were expressed, 11 wages decided in less than 45 minutes; the result is that everyone is happy! References have changed, we do not know how to engage employee attention toward the criteria, 2,700 to 4,500 euros of wages. Everyone chooses consciously and the collective respects that! No negotiation takes place.

Not only does this type of approach impose value creation but also a culture of giving and sharing. This is why firms have created the corporate hackers[13] movement. Relying on communities, "it starts, it is created and it spreads". The corporate hackers movement is here to show that it is possible!

Adam Kahane[14] has suggested two ways of moving forward: power and love. Each of these ways has a positive side and, if we abuse, a darker side, which must be illuminated in order to inhibit the negative side. These two ways can equally apply to the individual, a group, an organization, a country.

Power is useful in terms of mobilization, energy and problem solving. On the other hand, if power is abused, it turns into a bulldozer, a destruction tool, a means for eliminating opponents. Love is the driving force of engagement and collaboration and facilitates the understanding of situations. Used in excess, it becomes manipulation (for love). It suffocates the actions.

---

13 For more information: http://corporatehackers.org.

14 These are the books by Adam M. Kahane, for more information, *Pouvoir et Amour, théorie et pratique des transformations sociétales*, et *Scénarios pour la transformation sociétale, pratique de coopération multi-acteurs. Action dégénérative ou de long terme.*

	Social transformation cycle	
	Before Power mode	After Love mode
Generative action	Affirmation Mobilization Constraint	Commitment Collaboration Grouping
Introductory response	Resistance Fighting against Polarization	Search for compromise Capitulation Fusion
Degenerative or long-term action	Setting up smoke clouds Defeatism Destruction	Manipulation Weakening Choking

Table 6.4. *The transformation cycle, according to Kahane*

According to Adam Kahane, each of us has preferences and uses privileged paths. Adam Kahane has suggested both ways in the best possible manner. For this, he suggested they should be in dynamic equilibrium, one after the other. He made an analogy with walking: walking a path too far would be like walking on bare feet! Because of this, it is better to use both paths, without merging them. He proposed strengthening the path where the individual, the State or the organization is weaker, without reducing the dominant path. He also considered that dynamics are necessary, since a static equilibrium is not possible.

What pushes power and degenerates it is the denial of love. What drives love to degenerate is the denial of power. Consultants are acting on cultural transformation at the service of the Accor group. Each member of the organization belongs to a different circle and thus becomes better suited to understand each other and avoid blockages.

6.13. Feeble signals for understanding evolution

DEFINITION. A feeble signal is a paradoxical (societal, demographic, technological, environmental, economic, psychological) piece of information, which – when reflected upon – has a major impact on the future of society and its organizations[15].

The concept of feeble signal is very important for detecting breakthroughs in society 5.0. A first type of analysis appears when referring to generations of individuals. Regarding Humanity 1.0 generation types, an even more decisive "revolution" may appear with Humanity 2.0.

6.14. The generations

Society 5.0 invites us to reconsider things under the light of the different generations.

6.14.1. *The Beta generation*

Generation Y contributed its tools, the smartphone and the tablet, to the workplace, the Alpha generation entered the Robot scene, and the Beta generation endorsed digital tools: the smartphone – or rather what it represents – has become a gesture[16].

6.14.2. *The more "ecological" consumption of new generations*

For a decade, food has been questioned. Beyond the synthetic food announced for 2020, and the fact that fresh products are preferred to preserved foods, prepared dishes are declining in Europe (according to a study by PERIscop). The horsemeat crisis that focused on frozen food in 2015 was revealing. The overall decline in the quality of food products, or rather their distance from natural food, especially for private-label brands, affected consumer perception. The regular alerts on obesity and diabetes and the aging of the population are not in opposition to the idea of freeing up time spent in the kitchen. Awareness of the fight against food waste has provoked other effects, including the rise of pre-dosed packs.

15 This definition has been used and suggested by Philippe Cahen in his books (see bibliographic references).
16 Read in the letter of feeble signs by Philippe Cahen.

Since 2012, the attendance to restaurants in Europe has been decreasing. The economic crisis and the terrorist fear are identified as important factors. Even fast food restaurants have seen the number of their visits drop and have seeked to bounce back by creating small places of restoration related to transit spaces and introduced simpler menus. "Fast good" has acted as a response to fast food restaurants, something described as the fast food of great chefs, such as Bocuse in Lyon, and more recently, the organic food trucks by Marc Veyrat, as well as high-quality hamburgers (Paul Bocuse and Marc Veyrat being very well-known chefs in France). Online orders are growing, either for take away or for home delivery. Restaurant customers are benefitting from new services.

6.14.3. *The middle-class generation*

While cities vote for modernity, the wide periphery (starting from 30 or 50 km) votes for the rejection of globalization. The middle classes with children are deserting the cities and moving to the peripheries, while the young or the old who can content themselves with small areas remain in the cities. With the imposition of social housing percentages in the heart of the cities, the other housing facilities are becoming more and more expensive for purchase or renting: therefore, the young couples with children are leaving the cities. This might explain Donald Trump's political votes, Brexit and the *Front national*.

In this New World, the world of globalization, the general rise in educational attainment is a neglected variable. These "7 worlds" tend to close up, to fall back on themselves. For example, South Korea, which used to be poorer than North Korea in 1953, is now a leader of new technologies, with firms like Samsung. Thanks to its universities, Singapore has become a technological center. The world is more educated. As a consequence, the western middle classes are blending into the global middle classes, of which those in the developed world are the largest ones. This has an impact on the purchasing power (connected with salaries) of middle classes. These are the ones who benefit the least from globalization. In addition, training only affects skills for a short time. Since the least trained are the least paid, it is necessary to encourage education and for the State to initiate training courses.

6.15. Skills and generations

Skills are an essential factor for positioning individuals in society 5.0. The "has been" exists, for referring to previous firms and to those that adapted to society. The notion of skill has been adapted to fit the organizational context, as if it were something like an addition of individual skills.

6.15.1. *The distinctive skills of a firm*

The definition of a distinctive skill should be established in relation to the core business of an organization.

DEFINITION. The "distinctive skills" of a firm belong to its know-how and, when used on a product/market pair, they differentiate the firm from its competitors, on a dimension deemed as important by customers (F. Webster Jr, *Industrial Marketing Strategy*, 1995).

The notion of distinctive skill has often been referred to as "competitive advantage". This term derives from the assumption that having an advantage over competition facilitates success. The foundation of marketing leads to product differentiation. But it is important to pay attention to the formulation of distinctive skills. This brings the activities or trades of the firm together.

6.15.2. *The history of Low and Less*

In the spring of 1982, the journalist François de Closet published *Toujours Plus!* at Grasset publishing house. This essay was devoted to the forms of neo-corporatism that were eroding the French society, despite the expectations that surrounded Mitterrand's arrival to power. This book became a bestseller, some say a social reality. In fact, 850,000 copies were sold in the standard version and as many in the pocket edition.

6.15.3. *The cashless generation*

The cashless generation, which is common in Africa, is developing in its own way in France and in Europe: the Nickel account for those who do not have a bank account, the Ipagoo account for the multi-country and cross-currency needs. In different cases, be them African or European, the question is to keep exchanges fluid and simple, which is even more worrying to banks.

6.15.4. *Changes in commercialization and in business*

Trade has been questioned for the last decade. Whether the point of sale, the catalog, e- or m-commerce, everything needs to be changed. China is the champion of e-commerce and the United States is closing shops by the thousands. 3D printers were the Christmas gift of 2014 and revolutionized the industry. Decreasing prices and multiplying square meters are not the solution. Everything is happening right now as it has been occurring every 30 years since 1930. In order to reshape trade, we need to reconsider consumption and purchases, including the Alpha generation.

The American brand Costco opened in Seville. Costco arrived in France in June 2017. The implementation of this new store in Paris reflected the will to disseminate this model although, according to its French CEO, the number of membership cards did not meet the expected quota.

So far, the openings of European shops in America or American shops in Europe have resulted in failures. Costco is the champion of large quantities. I particularly remember that Costco manages one reference every 4 m^2 (13,750 m^2 and 3,500 references), while European hard discounters (Lidl and Aldi) handle a reference of 1.1 per m^2, 4 references for soft-discounts and a minimum of 12 references per m^2 are managed by a hypermarket. The choice of small choices is clever in terms of management, but this not necessarily astute from the point of view of the customer experience.

6.15.5. *Changes in the market*

Nowadays, the notions of market and that of competition seem to be distorted. In fact, new Internet or IT tools have been strong enough so as to change the boundaries and contours of markets and competition. Carpool sites are currently changing the contours of the transport market, in the same way that residential booking sites such as Airbnb are changing the real estate market and the hotel market.

So, the question is now to understand how these sites are positioned.

The other question involves the role of the State, which should make a distinction between the professional and the amateur, because taxation is fundamentally different between households and businesses.

If a member of Airbnb, Blablacar or Uber offers its services on these platforms and turns the project into an important source of income, what should the State do? Starting from which amount of money should we call him a professional? This gray work, this informal work, is it a service or a contribution made against a monetary exchange? Another question: should this be declared by the service provider member or by the platform? For economists, the role of money is called into question when we consider it beyond its level as a medium of exchange.

7

The Consequences of the End of Major Innovations

Since the 1990s, international organizations like the OECD have been writing reports about the end of innovation. The end of major innovations could have originated due to the fact that it is more complicated to meet the criteria of the society 5.0, which are sustainability, inclusion, efficiency and the power of intelligence.

7.1. The end of the major innovations: some observations

The innovations of the 19th and 20th Centuries mainly concerned technological advances. On the other hand, the innovations of the beginning of the 21st Century were characterized less by technical aspects than by the use of pairs such as objects/services or the "servitization" process.

The distinction that we study later concerning the nuance between incremental and major innovation (or breakthrough innovation) is still valid, but the facts show that lately we have not witnessed the appearance of major innovations.

Telephone	1861	Floppy disk	1950
Electric bulb	1883	Compact disc	1979
TV	1929	World Wide Web	1991
Atomic bomb	1945	Cell phone	1992
Computer (1st gen.)	1946		

Box 7.1. *Some major recent innovations*

The other feature of these innovations is that they are quickly occupying managerial areas and that they are developing faster. The evolution of innovation can be represented in Figure 7.1.

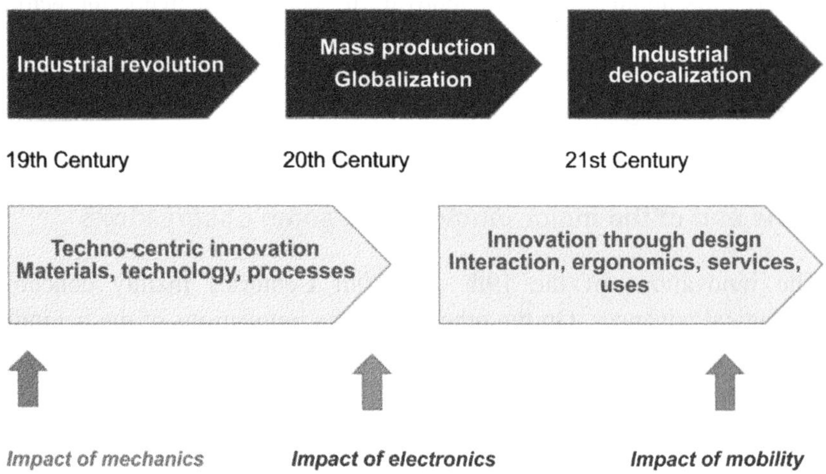

Figure 7.1. *Evolution of innovation. For a color version of the figure, see www.iste.co.uk/salgues/society.zip*

The innovations of the 19th and 20th Centuries were mainly based on technological advances. More recently, innovations have become less characteristic for their technical aspects than those related to uses. This has resulted in the idea of innovations through use and changes in production methods.

> "France has fallen behind in the field of innovation. [France] has no culture of use. For decades, our firms, mainly those technology-centered, have been oriented towards national aid programs for improving innovation and have consequently forgotten "user experience", on which really depends the success of innovation.
>
> However is the design that establishes the connection, which acts as the gear between technology and use, and which adds the share of imagination which makes objects desirable. Regardless of whether this technology is new or superficial, It is evident that France does not possess a design culture" [1].

7.2. Marketing philosophy as a vehicle for enhancing technology

Innovation will become increasingly related to a philosophy of marketing, itself derived from a new form of balance.

7.2.1. *Why do we mention a marketing philosophy?*

Marketing philosophy is based on the assumption that technology rules over the market (technology driving). Technology has an influence on the way of life.

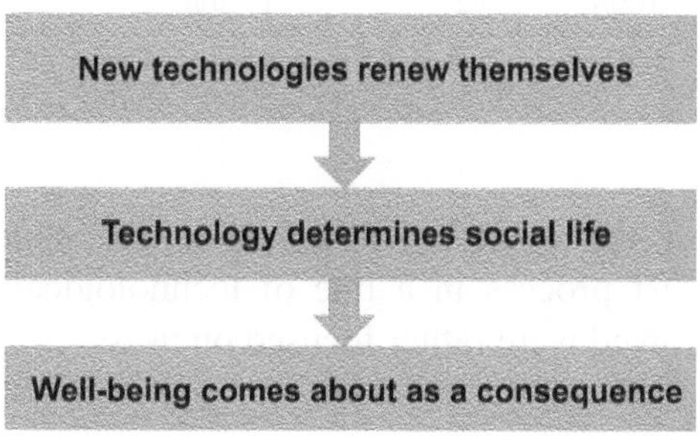

Figure 7.2. *Determination of well-being through technology*

1 Alain Cadiz, "For a national design policy", bill delivered to the Minister of Productive Recovery and to the Minister of Culture and Communication, October 2013.

Different marketing policies are based on this approach. One of the consequences is the constant renewal of technology. This renewal has a motto: the reduction of prices taken into account.

7.2.2. *The example of Intel processors*

The American firm Intel entered this logic of technology renewal. The price was certainly significant. Microprocessor 80486 used to cost 1,865 francs in January 1992, 727 in 1993 and only a few francs in 1996. On these bases, the microprocessor had an influence on the computer, which itself accelerated socioeconomic evolution. One of the conditions for this was to value the possibilities that computer systems could provide in terms of performance (mainly on productivity) and creativity.

Microprocessor	Year	Size in bits	Instruction number in Mips	Number of equivalent transistors
4004	1971	4	0,06	2,300
8080	1974	8	N/A	5,000
8088	1974	16	N/A	5,500
8086	1978	8/16	0.75	29,000
80 286	1982	16	2.6	134,000
80 386	1985	32	6	275,000
80 486	1989	32	20	120,000
Pentium	1993	64	112	310,000
Pentium Pro	1996	64	300	550,000

Table 7.1. *Evolution of Intel chips*

The problem of the evolution analyzed in the table is the fact that Pentium was the last process in a line of technological development; the processors that followed were rather focused on uses.

7.2.3. *Innovation balance*

The paradigm of innovation is that of a complex balance between viability, feasibility and desirability. This balance can be achieved in different ways: it can be due to the user, the uses or the societal context, or it can be due to changes in regulations.

One of the best-known examples is that of the radiators at the exterior of restaurant terraces, which are related to the ban on smoking in public places. This change in social order relies on a fully mature technology. Another example, bike rental through terminals in the city, is related to localization technology and the implementation of terminals. Saint-Gobain heated glass made it possible to build transparent radiators. Carbon fiber composites are useful for reducing the weight of technical bikes.

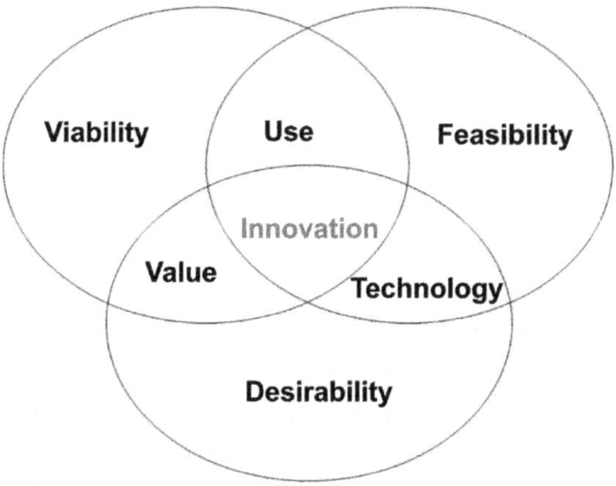

Figure 7.3. *Innovation balance. For a color version of the figure, see www.iste.co.uk/salgues/society.zip*

The three keywords are desirability, viability and feasibility. These call into question the acceptability stemming from a change in incremental innovation.

7.3. The new forms of innovation

In the face of the end of innovations and the social determinism of innovations, the OECD became interested in new forms of innovation which emerged with the advent of society 5.0, that is to say, social innovation and frugal innovation. The latter is often associated with the idea of the Low society or to that of the Less society.

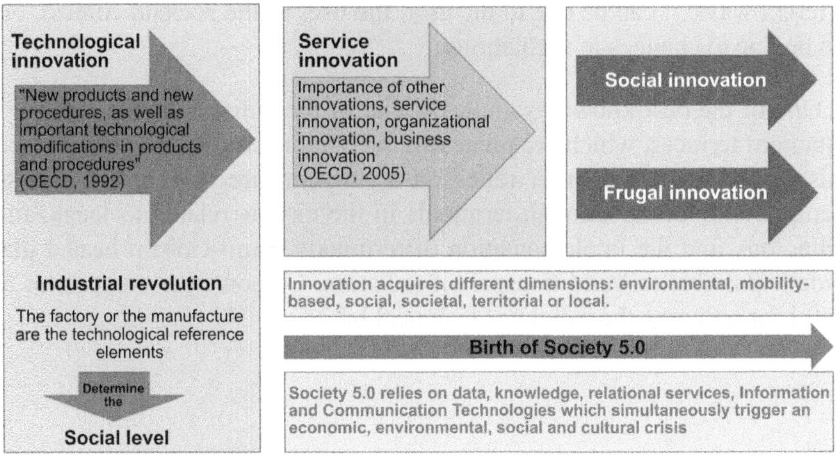

Figure 7.4. *New forms of innovation. For a color version of the figure, see www.iste.co.uk/salgues/society.zip*

7.4. The globalization of research

The globalization of research is the topic that several Science, Research and Society Forums[2] have decided to discuss. In concrete, the debate has not attempted a definition of globalization, which surprised me, since the topic is still controversial[3]. Round tables took the inexistence of globalization as point of departure.

7.4.1. *The globalization of science does not really exist...*

According to demographer Hervé Le Bras, both education and science did not receive a big impact from globalization, or at least, not so big as it seems. The author relied on various examples. He explained that disciplines vary greatly from country to country, at least in what concerns definitions and their approach. For him, the enemy that led to this absurd situation was the defense of "cultural relativism".

2 These forums, which took place in 2015 and 2016, were organized by the *Conservatoire des arts et métiers* and *Le Monde* newspaper.

3 In my book *30 ans de téléphonie mobile*, I draw a distinction between globalization and internationalization.

The pressure of global competition is not a reality. The argument that is often heard: "If I don't do it, others will!" is often perceived as unbelievable. The formulation of research is local in France, not to mention the examples of regional research. In fact, there are few global research topics, because societal needs are very different from one point of the globe to another. Only methods can be shared. For example, the industrialization of agriculture and 1000-cow farms are empty words for people who need to feed, to overcome drought and gain access to water!

7.4.2. *Scientific globalization is only real for mathematics, physics and health*

At least, this is the thesis advanced by Jean-Pierre Bourguignon regarding mathematics. For him, math skills have become strategic for businesses. The number of Mathematics engineers might increase by 10–20% in the near future.

According to Frédéric Bordry, the same will happen with physics. The general acceleration of society is in stark contrast with the long time constant research involved in physics. The project on the boson dates from the beginning of 1983; as a result of increasing costs, the returns of 2010 research and the means for carrying out research cannot help but be globalized. Thierry Bordry recalls that the LHC accelerator costs 5 billion Euros. Truth be told, these are amounts that may stimulate the industry, which can reuse the developments obtained. However, with such costs, research can only be global. Now, the main question is where the next big accelerator will be placed.

Health is also global. Cynthia Fleury recalls that viruses, bacteria and pollution are not stopped by borders. She incorporates a nuance, explaining that risk exposure differs from one geographical area to another.

7.4.3. *The key point is European research*

All the protagonists of research (or almost all of them) agree: confronted against major nations such as the United States, China and the Russia, research must be international, that is to say, European in what concerns us.

Cynthia Fleury intends to bring forward a model of European research. This is not only a need, but an integration element and a source of funding. While the European research program dates back to 1951, science and research were included under European skills, unlike education which has not achieved this status yet (it is via harmonization that Europe is competent in terms of education). Europe has still great work to do: Galileo, Boson, Airbus and Ariane are the few and only major European projects known at present. There could be others. According to Cynthia Fleury, it is necessary to privilege European research, more than other types of research. "When will we have a new Frankfurt School?" she suggested the thought-provoking question.

7.5. The globalization of scientific publications

The topic derived from the debate on science is that of research production and, therefore, of publications. Certainly, there is a globalization of publications, but in fact, the readers of these publications are not numerous.

7.5.1. *Scientific communication: publish or perish*

According to the actors of research, real globalization involves the mode "publish or perish". According to everyone, this will be generated by bureaucrats and politicians who have polluted researchers in every field.

After its launch in 2012, thousands of researchers, many academic societies and some publishing houses signed the San Francisco Declaration on Research[4]. Its intention was to limit the use of the impact factor in the assessment of research.

As often as he can, Frédéric Bordry quotes an example: Professor Higgs, who worked on the Higgs boson, has an H-index that would prevent him from being University Professor in the majority of the most popular establishments.

---

4 San Francisco Declaration On Research Assessment (DORA) was issued by the American Society for Cell Biology (ASCB) on December 16, 2012.

The pressure to publish is causing real ethical problems. The current situation features a fragmentation of publications which is reflected in an explosion in the number of scientific articles and a reduction in the number of founding or major publications. The result is that access to knowledge becomes slower. Bernadette Murgue has given some details of this: the researcher publishes the same paper with minor variations several times, on an average of 10 copies, which results in a lack of originality. Besides, in order to continue or to check research, it is necessary to "make a merger" of different journal articles, which turns the activity into something more complex.

7.5.2. *The solution, to expand the scope of "publications"*

For scientists, a solution could be to expand the notion of publication, which so far has remained limited to the world of so-called "scientific" publications.

For Jean-Pierre Bourguignon, it is difficult to take into account scientific differences and different forms of publication. Other forms exist: symposia, books, patents, etc. It is necessary to be aware of both the variety of disciplines and the modes of publication.

According to Cynthia Fleury, it is a good idea to remain attentive to Open Science movements, as well as the publication of raw results. Both of these allow the audit of research. Debates can arise from the use of these results. Counter-offensives and controversies may then appear. These "drawbacks" pose the problem of the impact of this approach on society, which may lead to new forms of research ethics. This ethics will contribute to the emergence of trust between science and society.

7.6. The role of bureaucracy in research

For Jean-Pierre Bourguignon, bureaucracy is the "disease". For him, the normality of control is initially invalid for research. Adapting modes of control to money practices will have to be dealt with by special policies. Doing research and becoming result-oriented according to a schedule do not make sense! The idea would be to obtain policies able to make control become adapted to the activity. Bernadette Murgue has insisted more on the role of barriers and various controls than on financial aspects. She has given some examples. In epidemiology, there is an unwritten rule according to which "a research project starts at the end of an epidemic and ends before the next". When you want to do research, the limiting factors are permission requests, some of which are related to ethical issues. However, these are requests that can take a long time in most cases.

Hervé Le Bras made a joke announcing that globalization has been faster for research bureaucracy of research than for research itself. Among the examples quoted, the National Agency for Research (NAR) has the same rules, regardless of the field of research. These were copied from abroad. Now, there only exist *a priori* controls, no more *a posteriori*. Regarding costs, these are no longer differentiated by discipline. For example, an amount of 150,000 euros may look ridiculous, but in social sciences, this is sometimes too much, while for other areas, it is not enough!

Hervé Le Bras illustrated with one example the excessive centralism that is particularly suffered in France. When asked for a report for the CNRS regarding the consequences of structural effects in the demography of researchers, he had to make 10 groups work together, for whom the objectives were announced on January 10, and the report for the central office had to be ready by February 22; all of this is to determine how many recruitments the agency would be able to perform. For Hervé Le Bras, it was a "grotesque bureaucracy" combined with a total "lack of strategic vision".

Finally, bureaucracy is amplified by the number of actors, including in the financing area. In China, there is only one operator! In Europe, the multiplication of actors is the rule.

7.7. The role of China

QUESTION. What is it that makes the Chinese become leaders in these fields?

To this question, Jean-Pierre Bourguignon provides a simple answer. Officially, bureaucracy is enormous, but, in reality, in France, there is great flexibility in terms of research. We should bear in mind that China has changed, China is not Beijing. If we pay attention to the nine leading universities from this country, all of their presidents have spent at least 10 years abroad. This, combined with a very large autonomy of the provinces, generates a very strong dynamic.

In China, a provincial Governor handles more than one billion Euros per year to be distributed among universities and private research. Besides, there is a real interest in developments proposed by engineers. Not only is the political decision-making context very different from the policy research of European countries, but in China, the share devoted to industry is more important and more oriented toward the long term. According to sources, China produces 150,000 scientific articles a year.

For Jean-Pierre Bourguignon, the contest between universities, a feature at the foundation of China's scientific milieu and Academy, is quite real. This could also be one of the elements of China's dynamism. To the question of whether China could work as an example, the panel replied that this was not to be discussed. Which might explain the appearance of the question: what is the solution?

7.8. The solution: to restore philosophy, poetry and morality to science and innovation

According to Hervé Le Bras, science, a phenomenon of great magnitude, used to be related to philosophy, but then became distant. For Cynthia Fleury, we are increasingly suggesting restoring humanity to science, something which is quite not the case! The author also mentioned that science should be careful not to step away from morality.

Restoring morality to science might favor the development of more risky research, or research whose results are uncertain. In virtue of current methods, researchers and their welcoming institutions are taking small risks in the choice of research topics. The consequence is that smaller amounts of daring research obtain funds.

The transformation of the lives of researchers is also a moral issue. Research is conducted by occasional workers! The precariousness of young researchers, which is the rule nowadays, is increasingly forcing them to get prepared for the "next job", instead of focusing on a long-term research axis. In addition, a strong psychological impact is observed. The generalization of precariousness has effects on the ambition of young researchers, who become involved in quiet[5] sustainability instead of scientific debate with an uncertain future. According to Cynthia Fleury, the limits of the reasonable

5 In this context, "quiet" refers to certain continuity, with no alterations.

have been trespassed. She added that a researcher cannot develop the same ambitions, depending on whether he can fit in a long-term professional horizon, as it used to be in the past, or in the precarious conditions of current research, which leads to the choice of opportunistic strategies.

7.9. The new research in society 5.0

Two new questions emerge.

QUESTION. How can we promote fundamental science, that is to say, unscheduled science?

Some researchers provide a simple answer to this question. It is only necessary that Selection Committees are not too conservative.

QUESTION. How can we value the contribution of researchers?

Consensus is being born among the actors of research on an idea. Measuring the impact of research on society would satisfy this appreciation. For this, it might be necessary to create larger evaluation committees, particularly extended to wider audiences other than peers. It might also be necessary to elicit more interest toward research activities on the part of the general public.

All things considered, many countries lack ambition and a national research strategy.

7.10. Innovation related to opportunities

We are in a period of social innovation, innovation by users, innovation by organizations or frugal innovation.

As many other countries belonging to the OECD, France has suffered a delay in the field of innovation. France does not have a culture of use. For decades, French firms, particularly technology-centered ones, have been oriented toward national aid programs for supporting innovation. All of these programs have forgotten "user experience", on which depends the success of innovation. They also have underestimated opportunities related to changes in societal conditions, changes that differ from one place to another.

However, it is design that makes the link, which acts as a gear between technology and use, and which adds the share of imagination that makes objects desirable. Regardless of the fact that this technology is new or incipient, it is nonetheless clear that France lacks a design culture (Cádiz, 2013). There are different sources of opportunity that should be explored, but these fields are generally not those proposed by the agencies responsible for funding research in Europe and quite rarely in other countries.

Opportunity	Example
Demographic changes	Population aging, golden age, silver economy
New knowledge	Health genetics
Combination of processes	Introduction to electronics and vehicle networks
Changes in the sector	The drug industry used to be a chemistry industry, it is now increasingly becoming an agro-food and a biology (birth of biomedicine) industry
Changes in the market	The determination to eat healthy food, rejection of pesticides and endocrine disrupters, choice of relations and short cycles
Changes in perceptions	Anxiety or boredom have introduced changes in perception, which act as a source of innovation
Unexpected events	For some manufacturers, terrorist attacks have prompted the birth of a security market. The same applies to networks and cybersecurity
Incongruities or bissociation	Bissociation is the association of two distant objects in order to make a new one. For example, the combination of the closet and cold became the refrigerator (1913).
Asynchrony	A gap between smart vehicles and the needed infrastructure

Table 7.2. *Examples of innovation opportunities*

The concept of incongruities or bissociation was forged by Koestler[6]. It exploded in the association of two concepts which are commonly unrelated. Other sources of opportunities can burst, thanks to switching, as a result of brutal changes.

6 Arthur Koestler, *Le Cri d'Archimède. La découverte de l'Art et l'art de la Découverte*, translated in French by Georges Fradier, Les Belles Lettres, Paris, 2011.

Type	Element	Example
Actor	Human feet + fish	Fish pedicure
Target	Gays and lesbians + matrimonial agency	Specialized marriage agencies
Material	Pedestrian + sensory sidewalk	Electricity production
Absence	Lack of resources + clothing	Emmaus movement
Experience	Tree + hotel	Tree houses
Facilitation	Gloves + smartphone screen touch	Special gloves for tactile screens

Table 7.3. *Bissociation innovation (Koestler)*

7.11. The paradigm of innovation

DEFINITION. Innovation is defined as the appearance of the product-service pair, achieving balance between viability, feasibility and desirability.

This balance can be achieved by accessing different paths or doorways, depending on the initial analysis of the context and the user. It can be reached provided that the walls that prevent their development do not prevent innovation from appearing.

According to the chosen path, different types of innovation may emerge.

Entry doors	Examples
Uses/Societal context	The radiators at the exterior of restaurant terraces are a reaction to a societal change (ban on smoking in public places), through the exploitation of a fully mature existing technology
Uses	Bike rental at city terminals
Technology	Saint-Gobain heated glass, enabling transparent or composite radiators, or carbon fiber for technical bikes
Desirability/Recognition/Appreciation	Radiator artwork for luxurious interiors or design bikes evoking a community, a universe

Table 7.4. *Examples of entry doors and product services*

7.12. Design thinking

Design thinking is seeking a compromise that will make it possible to gain competitive advantage. It is a balance between logic and intuition, or even marketing philosophy. The design phase is the result

– of logical or analytical reasoning: a controlled and a controllable result;

– of intuitive thinking.

The idea is to obtain a product or a service that meets use goals. For this, it is not enough to determine what the product or service should be. It is also necessary to imagine what it could be. The same function for creating value could be achieved, thanks to a different product or a service.

In industry, a competitive design process involves all the actors in parallel. The phases of creation, production and distribution are feature co-creation steps, a characteristic of "Co" society. In this approach, every player in the value chain works separately from others and should coordinate achievements. Creating a collaborative fertile ground for these design teams is often necessary.

DEFINITION. "Design thinking is a discipline that employs sensitivity, designer tools and methods which help interdisciplinary teams to innovate by meeting user expectation, technological feasibility and economic sustainability"[7].

The methodological foundations of design thinking are a smart mix of existing methods. This approach resembles the creativity steps proposed by General Electric in the 1980s. Design thinking approach can be summarized by different milestones.

7.12.1. *Stage 1: identifying a problem and understanding its environment, "observation phase"*

This inspiration phase questions the problem from a different perspective. Actors observe, analyze and synthesize by expanding exploration to

7 This is the design thinking definition provided by Tim Brown.

related fields, through the use of analogies. The idea is to understand the problem, and then to redefine it. In this phase, participants question etymology, definitions and uses. Arts, humanities and social sciences, business and technologies are implemented.

Observation refers to "watching what people do and do not do, and to listen to what they keep in silence". It is necessary to define an environment. Sometimes, it is useful to show empathy, in order to obtain additional information.

7.12.2. *Stage 2: finding the concept or idea that will make it possible to find a solution, "ideation" phase*

Ideation is a conceptualization phase which helps generate ideas, evaluate opportunities and choose axes to be developed. This phase concerns the design of the form which will embody this concept.

"Design & build" are the two following phases.

7.12.3. *Stage 3: designing*

The design phase requires identifying tasks and setting the locks. This phase requires the definition of users, uses, the context to be associated with feasibility and the economic aspects involved.

The tools used are benchmarking, story-telling and canvas.

7.12.4. *Stage 4: building a model and a prototype*

The building phase refers to defining models or prototypes. For this, the 5 W method is a possible approach.

Question	Sub-questions	Examples
WHO?	Who, with whom, on behalf of whom...	Responsible, actor, subject, target...
WHAT?	What, with what, in relation to what...	Tool, object, result, goal...
WHERE?	Where, from where, toward where...	Location, service...
WHEN?	From when, for how long, within what time frame...	Dates, periodicity, duration...
HOW?	In what ways, under what conditions, through which procedure...	Process, technique, action, average material...
HOW MUCH?	To what extent, values involved, at what dose...	Quantities involved, budget...
WHY?	Cause, triggering factor	Justification and causes which led to... (the "reason" behind, beliefs)
WHAT FOR?	Reason, purpose, objective	Justification by desire, ambition, forecasting...

Table 7.5. *The 5Ws*

7.12.5. *Stage 5: the assessment phase or "evaluation"*

In order to test the concept with users: the "evaluation" phase is built for this purpose. In this phase, the actors experiment and use the prototype that will be associated with tests. Different series of prototypes are carried out. The first prototype is a scalable object, the first one is very simple and the last one will (almost) be the finished product. Depending on its progress, it will be used for showing, explaining, testing, convincing or even making a presentation. Prototyping is limited to the product or to the physical prototype, it is the visual representation of a project.

During this phase, the respect for desirability, feasibility and sustainability factors is validated.

The implementation phase, ends up, values and deploys.

In practice, working with design thinking can be organized around three logics:

– the logics of co-creation, related to “Co” society;

– an intellectual approach alternating intuition and analysis phases;

– a field study (situational, with direct observation, etc.) complementing (or replacing) quantitative studies.

This method uses opening or closing logical approaches. The method implies divergence, so as to create choices (creativity) and then convergence (for implementing the aforementioned choices). Several convergence and divergence iterations are necessary, each of the iterations being more detailed than the previous one.

7.13. The risks of innovation

Innovating involves risks that have been studied by many authors. These risks can be classified according to different types.

Type of risk	Observation
Executive	Can we develop, produce, distribute the new product/service? Can support functions be deployed as expected?
Technological	Will the performance of the product/service satisfy consumer expectation?
Marketing	Acceptance of the product/service in price, quantity and disseminating period on the market
Ecosystemic	Need for co-innovation, for example, smartphones and apps
String of adoption	This risk appears, for example, when other actors are necessary for the use chain of a product[8]

Table 7.6. *The risks of innovation*

8 The elements that make up a car cannot be conceived without pondering the role of garage owners.

The risk in the adoption chain is a new risk, also related to ecosystems. The most commonly quoted example is the tire offered by Michelin which made it possible to drive for a long time without being obliged to fix it, but elicited the use of specific equipment by experts, which they refused.

7.14. The lessons of Thomas Edison

Innovation involves making a first strategic decision, which begins by understanding what type of business we should be in the future. This leads to three fundamental questions:

– How do we create value?

– How can we be of use to the customer?

– Which are the communication and lobbying actions that need to be implemented?

This often leads to the need to change the business model. Thomas Edison said: "Electricity is not energy, it is a means for transporting energy...". This means that the important invention is not the light bulb, but the creation of power stations, power lines to carry electricity and lobbying for encouraging the adoption of electric lighting, rather than gas lighting in the cities. This example is also interesting, in that urban innovation has also become an innovation of the home. It is easy to understand that the innovations of the smart city will spread within homes. That is why the smart city is a key point of society 5.0.

The second decision is related to Michael Porter's five forces. For Edison, it was the choice of carbon instead of platinum for light bulb filaments. This was cheaper and easier to implement.

Then, it is necessary to understand that economy is stronger than technology, and that only first arrivers win. This is the example of patent defense. Joseph Wilson Swan was the first inventor of the light bulb. Edison realized that electricity, as an innovation, competed against gas. In addition, he had the resources to sue Swan over violations of the patent owned by Edison. When an innovation turns out to be great, it is the economy that prevails, as it also happens with recent innovations. Uber, a chauffeured car rental firm, is making huge losses, but maintains low-price services in order to gain larger market shares. Despite its financial deficit, it cannot be accused of selling at a loss, because it generally sells the service at a higher price than the one paid to its drivers. The same applies to chain with a drive deposit.

The fourth decision is to generate excitement, buzz or to use viral marketing. Edison made public demonstrations of his inventions in order to convince the public and financiers of what he needed and to make them wait. Major players of innovation technologies make an announcement of their products, even if they abandon them later. For this, they showcase their products or leave some demonstration units to companies, hoping to create an ecosystem.

The fifth action is to produce innovations that become standards and platforms, instead of mere products or processes. We should not hesitate to allow competitors to have access to these standards, as well as to multiple sources of the product or service. The battle of the standards – VHS versus Beta, HDDVD versus Blue Ray – did not make it possible to develop these products calmly. This is currently an issue for the smart car.

Being a skilled manager is a fundamental competence for innovation. If we lack this "sense", which cannot be learned, it is necessary to ask for help in this direction. Thus, Edison filed a blocking patent (picket patenting) to protect his main patent. Skill also concerns the management of teams in charge of innovation, as well as agility in the management of business models. This agility is to imagine when and how a product will become obsolete. Currently, Light Emitting Diodes (LED) are replacing the traditional bulb by using fairly old technology for a new use, according to Koestler's bissociation principle.

As a lighting source, the LED typically illustrates the search for sustainability of society 5.0.

7.15. Methods for innovating

Methodology for launching a product or a service is needed in society 5.0. These are based on a set of tools that use data.

7.15.1. *The preliminary questions related to the genesis of a product or a service*

These preliminary questions require agreement on rules and using the maximum amount of available information.

At this stage, it is necessary to agree on or to accept certain points. The first acceptance is to know who will be the final decision maker for the launch or the characteristics of the decision-making circuit. At this first stage, it is necessary to imagine a provisional business plan and to have ideas on the business model which could work.

The product or service must be compared to existing ones, either identical ones among competitors or internally similar ones. At this point, it is necessary to use data and forecasting methods for detecting the latent needs of customers or appropriating already existing needs. Then, it is necessary to make a decision regarding the impact of reliability on quality and costs. In this case, we speak of necessary reliability, whose corollary is the obsolescence of the product, or sometimes results from a decision regarding forced obsolescence.

Different methods are possible and are represented by the prospective diamond.

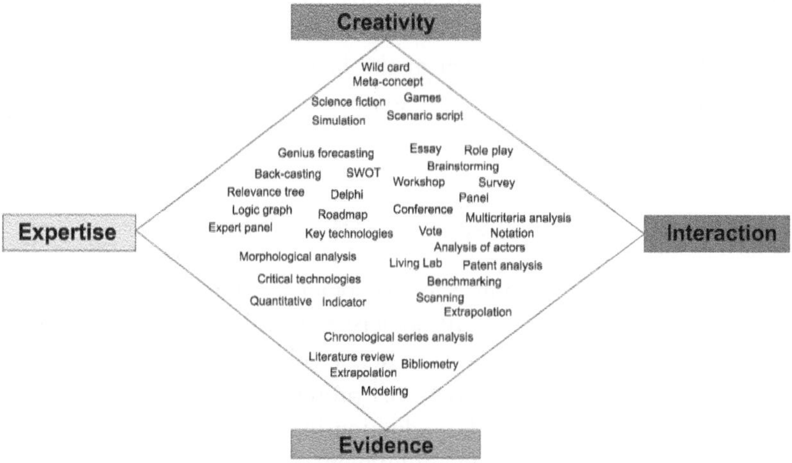

Figure 7.5. *The prospective diamond. For a color version of the figure, see www.iste.co.uk/salgues/society.zip*

The advantage of these methods is that they are integrated into software which are easy to implement. The downside is that these methods and associated data collection are difficult to use in small businesses because of the necessary personnel and the costs of licenses of applications.

7.15.2. *The choice on whether to innovate a product-service or to innovate a process*

This choice is critical in society 5.0. It is the source of a rupture or a paradigm shift in some areas.

7.16. Man in innovation

Modern man does not differ greatly from that of hundreds of years ago, except for the accumulation of knowledge and of know-how.

7.16.1. *The human resources of the innovative firm*

In many firms, human relationships are strongly associated with a corporate culture. While this relationship remains highly present in family businesses, the association of culture construction and human relations is considered as the magic potion of start-ups. Large companies are also trying to maintain this form of culture. The balance between the digital world and the search for the human element at the heart of the firm is necessary for businesses to stay innovative. This topic is broad and could itself be the subject of a book.

7.16.2. *The answer to the society of boredom*

Society 5.0 is also a response to the society of boredom. The society of boredom is one of the features of society 5.0. The boredom component of society 5.0 became possible as a consequence of digitization (information technology 2.0) and new modes of production (industry 4.0), which themselves have an impact in the reduction in the necessary working hours, thus increasing free time and unemployment. The problem comes from the fact that boredom is everywhere!

7.17. The different forms of boredom

Milan Kundera distinguishes three categories of boredom. Passive boredom is the one that appears when we yawn due to a lack of interest. Active boredom is the one that is present when we practice a sport or do something cultural, so as to spend leisure time. Revolt boredom is the one that takes place in the suburbs, when cars are set on fire or broken.

Gustave Flaubert used to observe a difference between ordinary boredom, related to a situation in which man evolves, and modern boredom, which is existential. The modern world introduced other forms of boredom, which some associate with the advent of industrial economy. Thus, Martin Doehllemann distinguishes four types of boredom. Situational boredom is characteristic of a situation in which someone is expecting another person, or waiting for a train or bus. Saturation boredom is the case where too much of the same is going on, and everything becomes commonplace. Existential boredom exists as soon as the soul is emptied. Creative boredom is characterized less by its content than by its outcome, when we feel constrained to do something new. We base our analysis on this typology of boredom.

Situational boredom is easy to spot, because it is exteriorized by yawns, by impatience. It often reveals a poverty of experience, a lack of "compensation object". In that case, the mobile phone, which enables you to call a relative, or to play while waiting for a bus or a subway, represents the "compensation object". Situational boredom is often an emotion.

Man is a big consumer of novelty or of meeting new people in order to escape the monotony of "always the same", which is a case of saturation boredom. In Greek, there is a verb for referring to idleness, a king of satiated and sickened character, but which does not perfectly correspond to boredom, it is *koros*. We adopt the term "koropathy" to define this disease, which is often treated by marketing.

Existential boredom is always linked to an emotional tone. It is a situation that we know, but boredom is born because of our presence[9]. In the case of these emotional tones, the future or the world appears as an infinite field of possibilities. Boredom appears when this field is reduced or limited, even to only one possibility.

In response to creative boredom, fashion has no other quality or functionality than the fact of being "new". The quality of an object depends on the type of this object, and an object without quality is an object without identity, which becomes quickly unusable. That is why branding is important in fashionable objects. Without branding, these are not recognizable, and their identity could disappear because they change all the time.

Type of boredom	Example of product providing an answer
Situational	Using mobile phone in public transport as game console to kill boredom
Saturation	Frequent change of taste of flavored yoghurts
Existential	Car or clothing, in the sense that they represent a social status, a form of identification
Creative	Gadgets that kill time, video games

Table 7.7. *Products in response to different forms of boredom*

7.18. The transgression phenomenon and the transcendence one

In order to discuss this topic, we can place the arguments at the various stages of Freud's analysis. It is never useless to define these two terms more precisely, since transgression and transcendence are simple, but insidious forms in response to boredom, which can be situated at the anal stage. In view of this, the market has all the features of the mother at the oral stage.

DEFINITION. Transgression means purely and simply an overrun, a breach of the limits. Moderate or radical, transgression remains at the same level. Binging on cake and drinking alcohol excessively are examples of transgression.

9 Waiting at the doctor's waiting room only becomes boring if you are yourself in the waiting room.

DEFINITION. Transcendence implies a qualitative leap, switching to something else.

Man, who attains a nearly religious experience at a music concert, in order to become an "addicted" musician, is close to transcendence. Electronic products, from the stereo to the computer, work as a source of true transcendence. We belong to the Macintosh Chapel, we only swear by Sony, etc.

Transgression and transcendence are elements at the heart of society 5.0. This is because we violate the usual rules of taxi companies, that Chauffeured Transport Vehicle (CTV) firms are born. Thus, Uber tackles the same market that taxis did at the beginning.

Liaudet has compared the role of the mother to the market. It suffices to simply rely on the latter, not to miss anything. It meets all your needs, all your demands. The market is a purveyor of satiety, it is nourishing[10]. At this stage, the authors only consider the demand and the need, and they do so by overlooking desire and expectation.

7.19. Boredom comes from the ugly

The ugly theory can be theoretically divided between the risk of uniformity and a lack of harmony.

7.19.1. *The risk of uniformity*

In the 17th Century, Antoine Houdar of la Motte, the author of fables, stated that boredom was born from uniformity. In an old book of 1853 devoted to the aesthetics of the ugly, Karl Rozenkranz described boredom as being ugly. It would be a trivial thing if this author did not support that boredom is an opening to the comic. By contrast, beauty might make us forget time. At the same time, Alfred de Vigny considered boredom a disease, and one of its remedies was to love beautiful things.

Nowadays, the forms of boredom are more due to the massification of production, one of whose answers is mass customization. It suffices to give

10 Jean-Claude Liaudet, *op. cit.*, p. 93.

the impression to consumers that they have access to customized objects. Many examples have emerged. Let us observe the example of Coca Cola: looking carefully, it is possible to find a bottle with one's name on it. This quest is additional work for the buyer who values[11] the object. We can include Nike sneakers, which are also customizable. For some brands of clothing, the firm re-stains the garment and affixes embroideries on the holes. Citroën has announced more than 3 million combinations of roof, wheels, bodywork and dashboards for its DS3. Confectioners are launching personalization services, thanks to food 3D printers installed at their stores. M & Ms has agreed to customize its candy with short messages or pictures.

7.19.2. *The search for harmony*

The classical period replicated the ideal of harmony to the social level, by making what is beautiful, pleasant. Thus, ugliness only became a break of harmony. Ugliness is also an imbalance, an asymmetry, bad taste. Marc Augé defined harmonic systems as situations where rules stemming from filiation and descendance coincide with the rules of residence and localization[12].

This approach corresponds to the sociological and anthropological visions introduced in marketing.

Harmony raises other questions. According to many authors, we live in a society of suspicion. That is why the notion of harmony arouses mistrust.[13] Vergely wondered whether he is right to think that we need harmony at all costs in the field of morality. Harmony does not entail merging, which is confusing. More likely, it refers to the resonance of two elements, two distinct personalities. The more we are distinct, the more chances we get of being in harmony. It is probably for this reason that differentiation in marketing is pushed to the extreme, so as to reach the famous "one-to-one marketing", which is differentiation pushed to its limits, and consequently, to harmony.

11 After launching maximum customization on demand, Coca Cola reviewed its ambitions and limited its production to the most popular 250 names per geographical zone, as a result of labeling extra costs.

12 Marc Augé, *Où est passé l'avenir ?*, p. 42, Panama, Paris, 2008.

13 Bertrand Vergely, *op. cit.*, p. 43.

7.20. The search for equilibrium

It is in the spirit of a creator that a product or a service is born. But translating this vision into reality is more complex. The new product or service is the result of a delicate and complex balance:

– between the creativity of individuals who are responsible for it;

– the agility of the organization;

– branding genius within the organization;

– business requirements related to the knowledge of the customer.

It is on the collection of data concerning this balance that lies the success of a product and a service in society 5.0.

7.21. Design as a technical answer

Design methodologies provide a technical response, both to desirability and to viability.

7.21.1. *Industrial aesthetics and design laws*

The basic concepts and fundamental problems related to design are explained in the Charter of industrial design and in Jean Viennot's definition (1953), who was the founder of the Institute of industrial design.

DEFINITION. Industrial aesthetics is the science of beauty in the field of industrial production. It comprises working places and environments, means of production, and products.

This definition does not take into account the society of service which characterizes society 5.0 and industry 4.0. The Charter of industrial design consists of laws that must be respected by the production of goods or services.

DEFINITION. Law of Economics: the idea is to engage the economy's interest in the means and materials employed, provided that this does not impair a functional value or the quality of the work. This law was placed at the first place in the Charter.

It is very important in the design of the production of society 5.0. It is easy to understand the importance of design of low-cost services, as they thrive especially in air transport.

DEFINITION. Law of functional value: the concept of industrial beauty necessarily implies that works are adapted to their functions.

In other words, the product or service must be manufactured to fulfill a function, even if, subsequently, it is the misuse of the product that dominates.

DEFINITION. Law of unity and composition: to form a harmonious totality, the various bodies must be designed in function of each other and of the whole.

DEFINITION. Law of harmony between appearance and use: there is never conflict, but always harmony between the aesthetic satisfaction felt by a disinterested spectator and the practical satisfaction provided to the one who uses the product or service.

DEFINITION. Law of style: the aesthetic character of a work must take into account the normal duration to which it needs to be adapted. A useful work can claim a character of lasting beauty if it was designed away from the influence of fashion. The aesthetic character of a period follows the style which constitutes its expression.

DEFINITION. Law of upgradeability and relativity: Industrial aesthetics does not present a definite character, but is constantly evolving (progress and technical development).

DEFINITION. Law of taste: industrial design is expressed in the structure, shape, and balance of proportions.

DEFINITION. Law of motion: those objects that were designed to move by themselves find that, when in movement, they achieve the essential feature of their aesthetic.

DEFINITION. Law of hierarchy or purpose: industrial design cannot ignore the purpose for which the objects are produced.

DEFINITION. Commercial law: industrial design finds its application on markets where the law of the largest number of buyers cannot overturn the values of industrial design laws.

DEFINITION. Law of probity: Industrial aesthetics involves honesty and sincerity in the choice of the materials used.

DEFINITION. Law of the involved arts: industrial design involves an integration of artistic thought in the structure of the creation.

7.21.2. *The evolution of design needs*

The sources of design have evolved since major innovations have been taken into account. Society 5.0 is reusing this activity linked to the industrial revolution, an activity related to production. It contributes with major changes due to the presence of the end user in the design process.

Before	In the case of society 5.0
Mass production	Products adapted to the customer
Products related to mass markets	Products related to market optimization
Activity related to the political, historical and technological context	Global product, beyond specific contexts
Activity related to successive games reacting to technology	Use of these technologies by considering them as existent

Table 7.8. *Evolution of the design within society 5.0*

The responses of society 5.0 to design may acquire several forms.

This may be the total and simple refusal of industrial civilization and the attempt to return to an imagined society: this could be, for example, the use of short circuits in feeding technologies. We might go one step further and achieve the dream of returning to the past, by reintroducing the horse as a means of transport. It can also be illustrated by a “retro” fashion, which is often only superficial.

The other way is to know the new possibilities offered by technology and to use the approach of the designer so as to implement new products and services, more in line with the expectations of society 5.0, an approach that we describe as "the median way".

The ultimate solution seems to be the achievement of rational beauty, which is often not very objective, but linked to promises, including that according to which the human body could be more beautiful, thanks to such diet or such slimming cream.

All in all, design can be conceived as an attitude for

– making the object communicate by its function;

– using new materials, which were not usable in the past;

– using machines in order to produce more functional forms.

In all these cases, design contributes to innovation, but is not its source. It is there to respond to boredom, particularly the one linked to the shape of products.

7.21.3. *The use of a former theoretical approach in design*

A theory of the form matched the taste of the day and became relevant, thanks to semiotics, a discipline that can be broken down according to a triple approach:

– styling;

– formalism;

– functionalism.

The form (Gestalt) theory is a theory that separates sensation from perception. This theory is a theory of the whole, where form is everything. Even when forms are incomplete, perception prevails (example of the triangle).

DEFINITION. Gestalt law or transposition law: a form can be transposed; when modifications do not alter the structure, the form remains recognizable.

Besides this, there is the figure-ground law.

DEFINITION. Figure-ground law: the figures become a reference in relation to the background. In a heterogeneous field, a figure appears as something that offers an outline and is differentiated from the background, offers privileged directions and is organized.

DEFINITION. Law of unity segregation: some units are spontaneously recognized. Orientation, distance or the similarity of elements make this possible.

DEFINITION. Law of the right shape: simple geometric forms are striking and impose themselves.

That is why in the times of early mobile phones, consumers could not imagine a phone without an antenna. That is why antennas were added to the phones, even if they were not functional.

DEFINITION. Prioritization law: we differentiate what is important from what is secondary with the help of stereotypes (verticality, etc.).

The basic theory makes a distinction between

– syntactics, represented by forms, colors, texture, flavors and the layout of elements;

– pragmatics, which represents the interest of the product or service and is often considered as its use value;

– semantics, which is made up of the cultures, habits and the expression of pleasurable fashion.

Semiology is a former contribution by Ferdinand de Saussure, from his general linguistics course (1916), by C.S. Peirce, in his general theory on signs and Roland Barthes, in his rhetorics of the image. The idea is to approach the notion of sign by breaking it down in three ways: design and its semantic dimension, formalism and its syntactic dimension and functionalism with its pragmatic dimension.

Design considers that what prevails is the symbolic value of the product. Drawing is then the most important aspect. Its purpose is to highlight the appearance of the product. Its role is to visually seduce the consumer.

With formalism, the formal layout of items takes precedence over other requirements such as industrial criteria, the important role of geometry and also the choice of simple forms (including circles, squares, etc.).

Functionalism was brought forward by Sullivan. Form follows function (1896). Products are determined by their functions. It is a way of condemning pure aesthetics. He rather encouraged implementing the ideas of rigor and discipline. The shape is the result of a multiplicity of functions, among which usefulness prevails.

It is through achieving the balance of these three approaches that products will be accepted by the consumers of society 5.0.

7.21.4. *The aesthetic components*

But the aesthetic components of the product or service cannot be rejected.

They are related to cultural factors, such as habits, knowledge, culture and religion. Aesthetics is relative to cultural factors. The Eastern perspective might correspond to spatial expansion, whereas the Chinese world focuses on the differences between that which is empty and that which is full (Yin, Yang or earth, sky). From this stems the idea that symbols can play an important role.

Historical factors form part of this aesthetic, but are essentially related to the context. Historical factors are determined by the relationship between objects and periods, such that the chamber pot became obsolete with the widespread use of toilets in the apartments. These factors play an important role in the different types of relationships to the object. The existence of technology itself is a historical factor, as the need for these objects. For example, the development of the electric car requires the provision of an object: recharging stations. Among the historical factors, it is necessary to consider the phenomenon of the technological path, described in previous chapters.

Technological factors manifest in the form of state-of-the-art contributions, their accessibility and their acceptability. Technological factors are related to the role of the technical culture of the moment. This is the example of iron, represented by the Atomium, a monument in Brussels which represents an atom of iron. These factors create a relationship between society, objects and other associated services. This is how phenomena cycles are created, objects transform society and society produces objects. These cycles are related to economics.

Social factors are based on values: look, identity, symbols, fashion and communication degrees. Social factors are closely related to the role of the symbolic function and the effects of fashion (Roland Barthes). In kitsch culture, the social value is heightened and the fake reigns supreme due to the "ersatz". Social factors can be analyzed thanks to tools such as lifestyles and the use of the sociometrics (Jacob Levy Moreno).

Harmonic factors should directly be linked to notions such as proportion, coherence, rhythm, structure, unity, and the laying out of elements. These are related to the syntactic dimension of the products. With this approach, form is not the sum of the parts, but a function comprising several variables (gestalt psychology). The harmonic relationship depends on the permanent choice of the basic elements that make up the product and service.

Functional factors comprise form, function, durability or obsolescence, adaptability, and intelligibility of the economy of the means employed. Functional factors are fundamental in the fit between form and function. There is a notion of intelligibility: the function and its perception by the user are important, but this results in the risk of banalizing the product. The notion of analogy is useful. Design builds formal analogies with elements which have the same function. Added to this is the notion of homology: there is a similarity of forms which is a formal inheritance.

7.21.5. *The impact of the sociometrics and homology*

These two approaches are quite different, but both have a social influence.

The term "sociometrics" was introduced by the Viennese Professor Franz Xaver Von Neumann-Spallart in 1887. Sociometrics is a science which makes it possible to measure the economic and social situation of populations. It is positioned between sociology, which is too qualitative, and political economy, which has difficulty in becoming interested in individuals and their relationships. According to Auguste Chirac, in 1905, the method was intended to highlight the reciprocal links and the close solidarity between economic and moral phenomena. The following stage intended to carry out forecasts. At the same time, psychiatrist Jacob Levy Moreno had brought them forward in an analysis of social networks and, more generally, in social psychology.

DEFINITION. Sociometrics is the study of the interpersonal relations of a group, at a certain moment and for a given situation.

The basic concept used in sociometrics is the social atom. This defines the individual, not as a single being, but as a network of relationships, of which he is the point of departure and the destination.

DEFINITION. Sociography makes it possible to represent the attraction and rejection relationships of individuals within a group.

Most importantly, Moreno perceived sociometrics and the sociogram as instruments for measuring actions from the quantitative point of view. The objective of sociography was to implement reorganizing actions. It also strived to make them effective. Nowadays, these are used for analyzing the relationships in social networks.

The term "homology" comes from evolutionary biology.

DEFINITION. Homology refers to an evolutionary link between two features (typically, anatomical) observed in two different species, which is due to the fact that both inherited it from a common ancestor.

These common features are called "homologous". They may be anatomical, molecular (protein) or genetic (DNA) features. In innovation, some elements are homologous, because they established themselves through time. The keyboard of our computers and current tablets have their origin in the typewriter. The railway track gauge corresponds to the spacing of carriage wheels of the same era. Homologies make it easier to accept important switchings imposed by the technological change, either by the user (secretaries using the first word processors) or by the productive forces (the railroad case). This is the use of the already-described technological path phenomenon.

7.22. The sources and forms of design

The will to design has its origins in the different functions of the organization. To design is to create new products, but not necessarily to produce innovation. One of the questions that arises in this case is the following: "where does the original idea come from?" In this operation, the presence of data provided by society 5.0 facilitates implementation. An important action is the mash-up. This notion will be detailed later.

Function in the firm	Examples
Direction	Knowledge of a niche to work on Idea or ideology
Research	Innovation or improvement coming from the research centers or purchasable at the exterior of the company
Competition	Variant or copy
Marketing	Identification of need Desire, expectation, demand Request for the evolution of an existing product
Financing	Forced need of obsolescence for reasons of income or exclusion of a competitor (to make the competitor appear as outdated)

Table 7.9. *Sources of design*

The genesis of the product depends on priorities, strategy, politics and the culture of the organization. Commonly, four forms are studied: market driven, product driven, technology driven, competition. In society 5.0, these forms still exist, but another one, related to collaboration, must be added.

Type of Genesis	Observations
Market driven	The market asks us for the product or the firm that studies the market in order to get to know this request
Product driven	Innovation is delivered after doing research throughout the four phases: identification of uses or potential uses, building a matrix between potential uses and potential customers, looking for segments and target, defining commercial products
Technology driven	Use of technological developments
Competition	Response to the attacks of the competitor
Buyer-producer cooperation	The buyer is involved in the definition of the product ("normal" case in Business to Business offers)
Cooperation between producers	Producers cooperate either to share costs, or by waiting, A produces in country X what B produces in country Y and *vice versa*

Table 7.10. *Sources of design and their different forms*

7.23. The other criteria for innovating a product or a service

These static objectives can be opposed to the usual questions. The notion of use criterion is regularly used for analyzing the relevance of a technological tool. These are acceptability, accessibility, learnability, usability and usefulness.

Term	Content
Acceptability	Is it a response to a need in my daily life?
Accessibility	Can I really use it?
Learnability	Can I try it and learn to use it?
Usability	Under what conditions can I use it?
Utility	What can I use it for?

Table 7.11. *Innovation criteria*

8

Innovation in Society 5.0

Innovation in society 5.0 is characterized by the changes in paradigms and in management systems. Innovation has changed. The debate between breakthrough innovation and incremental innovation no longer makes sense. Society 5.0 is characterized by

– innovations in service, or "servitization", which is the classical kind of social innovation;

– open innovation, which is shared, in order to favor its development, generally by means of platforms, and which is quite a peculiar kind of social innovation;

– frugal innovation, which is characterized by achieving more with fewer resources.

In all cases, these innovations correspond to Koestler's writings. They are related to the "bisociative act". It is the innovative bond that suddenly connects reference systems which had remained separate until then. This bond makes us see or understand reality on several planes at the same time, which may assume several paths, even traditional ones.

8.1. The innovative product service

We can also consider the stages that exist from the creation of a product idea and its commercialization, its subsequent acceptance by the customer, as well as the consequences of these various levels.

8.1.1. *Losses during the innovation process*

Throughout the passage from the original idea to the product or service, losses are important indeed. Different studies have evaluated the loss coefficient.

Figure 8.1 shows an example of this.

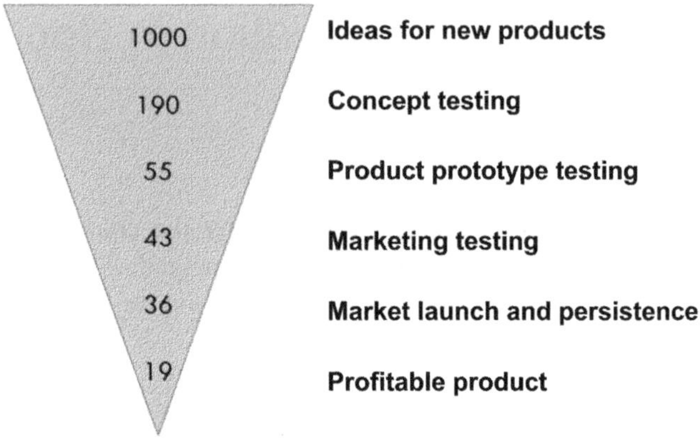

Figure 8.1. *From the idea to the product or service*

This calculation is one among many possible examples; it yields a ratio of 1 in 50. Often, there is confusion between an idea and the notion of a product. For instance, the pharmaceutical industry often assimilates a new validated molecule to a drug and complains about the failure rate of its sector, which, in fact, is not abnormal at all. The same goes for start-ups created in incubators.

In his course, Jacques Baronet, from the University of Sherbrooke, uses Figure 8.2, resulting from his research activity on a group of firms.

More pessimistic than the previous example, this curve yields a ratio of one in 3,000.

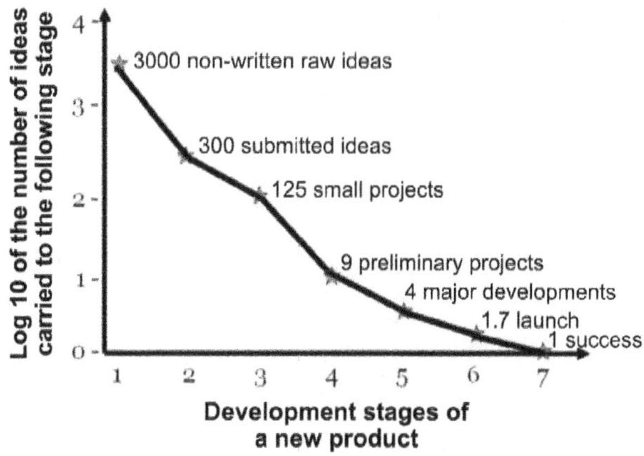

Figure 8.2. *Evolution from a raw idea to success (according to J. Baronet)*

8.1.2. *The question on the validation of a new product or a service*

A new product is validated when it brings a reply to three questions.

First of all, it should be coherent with the general strategy of the firm or the organization. Besides, a new product should respond to one of the purposes of the regime (efficiency) or to a financial goal (profitability, financial return, productivity). Finally, the new product should be placed in harmony with the marketing or management stages. The questions we consider regarding the products are specific.

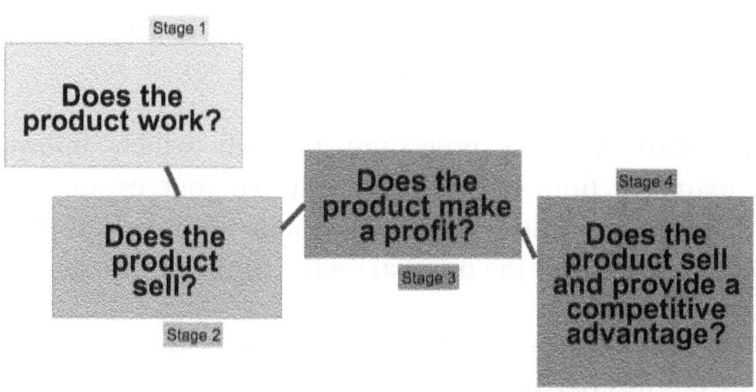

Figure 8.3. *Evolution of a product's stages. For a color version of the figure, see www.iste.co.uk/salgues/society.zip*

To these four questions, we should add that in society 5.0, a new question emerges: it is necessary to find on which data it is possible to rely and how this information may be collected and analyzed.

8.1.3. *Improving a product*

We are often driven to improve an existing product or a service, depending on

– the evolution of demand (consumer changes, changes in purchasing power, fashion effects);

– changes in the way economy works (shifting from an effectiveness regime to an efficiency regime or the other way around).

– the evolution of supply, particularly due to changes in the competition;

– new possibilities for collecting data.

Similarly, financial parameters (gains in productivity, financial return or profitability), seasonal phenomena, the effects of obsolescence or the will to redirect corporate identity may appear as important reasons for changing/improving a product.

8.2. The paradigm shift

Society 5.0 is characterized by important changes regarding innovation. Sustainable innovation actually reflects numerous disruptions. Innovation based on a linear evolution of scientific knowledge is no longer prominent. That is why sustainable innovation is either social or frugal. It is becoming more and more related to the concept of mash-up.

DEFINITION. Mash-up is a procedure through which information and knowledge extracted from one or more external sources are aggregated or reprocessed. Mash-up actions generally obtain information through Open Data, from the publisher or the original owner of the information.

In order to implement mash-up, it is necessary to have enough stock of open data in the desired field, and hence the importance of open data for setting up society 5.0. Networks are one of the sources of open data which have an influence on the performance of organizations. At this point, it is convenient to make a distinction between the notion of accessible data and that of free data. Free data are accessible and usable by others, which means that they are not especially protected. For this, there are different levels of accessibility.

Innovative questioning seeks to create novelty:

– by implementing user-based innovations;

– by managing open innovation;

– through the implementation of systematic innovation processes;

– by creating experimental mash-ups (or associations).

It is difficult to predict the future, because the innovation process is volatile, uncertain, complex and ambiguous. Society 5.0 is a "vocation society", in the sense that nowadays, we see that major innovations are closely linked to new organizations, instead of changes being applied to existing products or services. Platform economy is an example of this. While it is true that carpooling was born in the 1970s via "dazibao" posters, later on, small magazine and newspaper ads, the Internet and its platforms widely disseminated and simplified this sharing economy service.

"Large deviations" make it possible to implement competition; these deviations transfer progress from one sector to another. Due to this, innovation depends on the point of view of the people who uses or benefits from it. Innovation requires connectivity, an open interaction with the real. It is not controllable, but there are ignition and catalyst effects between interconnected systems.

DEFINITION. Ignition is the process whereby an idea, a product, or a service is "lit", or brought forward at a certain moment.

DEFINITION. Catalysis is a process referring to the acceleration, reorientation or selectivity of the kinetics of an action, by using a "catalyst" agent.

Catalysis comes from the ancient Greek κατάλυσις, *katalysis*. Selectivity makes it possible to steer reaction in a privileged way, in such a way that a product, rather than another one, is created and disseminated. Among catalysts, we can mention living labs, Fab Labs and incubators.

Society 5.0 needs new spaces of innovation where connections play a decisive role. That is why innovative people are less and less related to "inventors", but

– to curators who manage knowledge and know-how libraries;

– to bridgers, that is to say, people who are able to create links and bridges. Bridgers operate following the logic of bisociation.

8.3. Mash-up forms

Mash-up takes up different forms, depending on the case. These forms are the consequence of the implementation of a new informational paradigm.

QUESTION. How do we access technologies in order to create new products?

We can mention a number of pathways to access new technologies:

– the internal creation of competence, whose disadvantage is that it is expensive and time-consuming;

– the acquisition of firms mastering the technology in question, even if this strategy is dangerous, mainly due to prohibitive costs and the difficulty for integrating new teams. By contrast, this approach is being implemented by large groups seeking to make a technological shift, for example, by buying start-ups.

– cooperation agreements, although it is necessary to preliminarily reach full agreement on all the terms of the contract and the means to be implemented. Some cooperation structures can also be implemented in order to achieve this goal;

– external research contracts are limited by the ulterior difficulty of new developments;

– the outright copy and adaptation of products or services, in order to circumvent intellectual protection measures, trademarks and patents.

These different pathways are all problematic because of the disadvantages that they involve. The only method that has a small number of drawbacks is the access to technology via the purchase of a license: the post-war "Japanese miracle" is the most obvious example of this. The Chinese have been accused of making copies, but they have subsequently implemented innovative industrialization processes, resorting to a mix of the abovementioned ingredients.

8.4. "Co" society

One of the major consequences of the evolution of innovation as it has just been described is the birth of "co" society, which characterizes society 5.0. This evolution will be analyzed in detail later on.

Innovation is a synergy. More than ever, major groups and start-ups are working together. First, this is achieved, thanks to a pooling of processes and tried and tested tools. The striking power of large firms is thus associated to the agility and adaptability of small ones. Co-construction and co-acceleration are interconnected. The increasing number of ignition places, often called accelerators, proves this. Large companies are accompanying their own digital transformation and developing dedicated products and services which they draw from young innovative firms.

8.5. The sharing of information

Innovation sharing is a necessity for implementing "co" society, as we have just described it. This sharing is multiple and may assume different ways.

In innovation teams, there should be a sufficient dose of information sharing, but not excessively. Too little communication creates clans, clubs and cliques. In addition, little communication isolates teams. Too much communication leads to group thinking. This means either adopting the vision of others, which is generally common and with no future, or choosing ideas created by competitors, as an obscuring smoke cloud, which finally places the competitor in a more favorable position. Everybody thinks the same, which ultimately decreases originality, and innovation, in the long term.

8.6. Social networks, Internet and innovation

Social networks and Internet websites are tools for accelerating business, aimed at company directors, managers, but also at retailers and sole traders. Social networks are both a key element for business strategy and communication, and a source of information. Several action types are required and described in alphabetical order in Table 8.1.

Name	Type of action
Ambassador of the brand on social media	Setting the framework and constraints of corporate communication, establishing explicit, simple and clear lines of communication of the organization and acquiring communication techniques adapted to social media
Community management	Managing existing communities or those that are created and have an impact on the organization
E-reputation	Monitoring and improving the reputation of the organization on the media, the Internet and social networks
Performance on social networks	Improving overall visibility on the web and on social networks, in order to improve the reputation of the organization (qualitative and quantitative visibility)
Search Engine Optimization (SEO)	Acquiring the best SEO practices and online reputation management
Competitive intelligence	Organizing and achieving an effective competitive intelligence on the Internet and on social media

Table 8.1. *Different types of action on networks*

8.7. The collaborative forms

Collaborative economy dates from around 2011. Today, it is present in our lives on a daily basis. The case of Leroy Merlin is interesting in more than one way. No. 2 DIY worldwide leader opened TechShop workshops (3D printer, hydraulic cutting, digital machines, etc.) in its points of sale. As it did with the 104, Leroy Merlin has switched to makers: if you have not understand “104” and “makers”, it means that Leroy Merlin is one step ahead. TechShop has a goal of rejuvenating the customer base. This goal will be clearly visible in less than 10 years’ time, but in the meantime, the firm has really gained an advantage.

8.8. Innovation ecosystems

When we cannot access bridgers, a demand is created at the level of ecosystems, and this is often represented as an "innovation melting pot", actually creating an ecosystem that favors innovation, thanks to its ingredients.

8.8.1. *Resource centers*

Resource centers are necessary elements for implementing innovation. These were created within society 4.0. Resource centers are sometimes called "competence centers".

DEFINITION. Resource centers are organizations which offer technological infrastructure and associated knowledge and skills for resizing and developing innovations.

The notion of a resource center was formalized by the European Commission, within the framework of the industry of the future initiative, focused on small- and medium-sized enterprises (SMEs). The initiative was called "ICT Innovation for Manufacturing SMEs" (I4MS)[1] and was launched in July 2013. The four main types of innovation were brought forward at that time:

– robotics;

– high performance simulation in the "cloud";

– applications based on laser and light;

– systems including smart sensors.

This initiative involves a list of competence centers offering solutions and technological infrastructure to SMEs. This initiative also intended to define a precise vocabulary for describing technological innovation. These skill centers were introduced as an effective wheel which could help innovation move forward.

1 For more information: http://www.i4ms.eu.

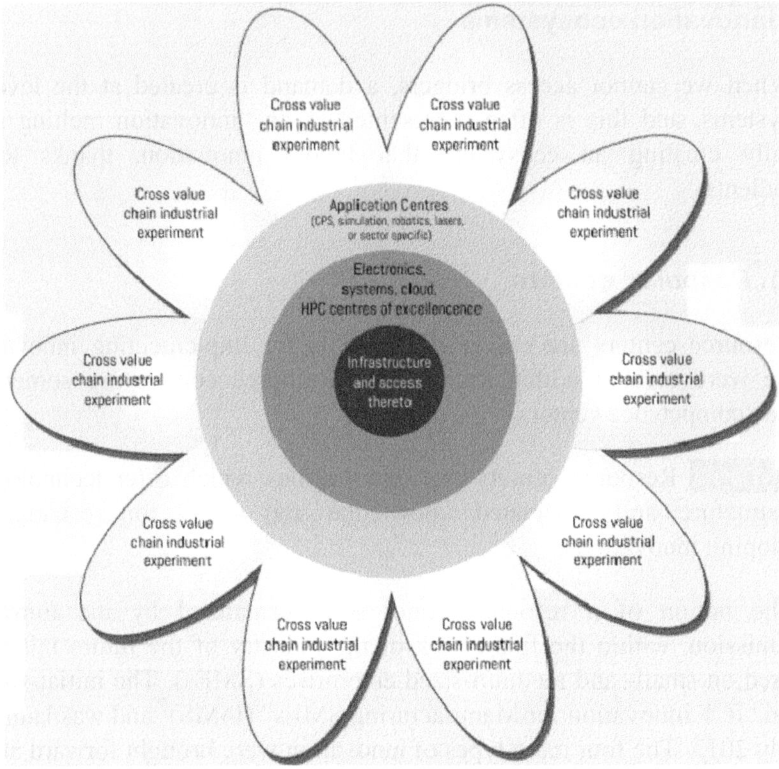

Figure 8.4. *Presentation of the innovation wheel (source: I4MS). For a color version of the figure, see www.iste.co.uk/salgues/society.zip*

8.8.2. *The concept of the Digital Innovation Hub*

Throughout 2017, the European Commission encouraged the emergence of Digital Innovation Hubs (DIHs).

It introduced DIHs as a method for digitizing industry. The reference diagram is illustrated in Figure 8.5.

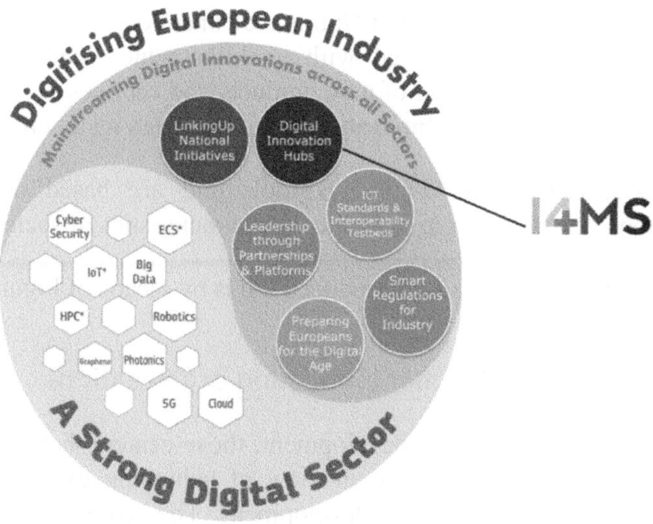

Figure 8.5. *Place of I4MS in the European Digital strategy (source: I4MS). For a color version of the figure, see www.iste.co.uk/salgues/society.zip*

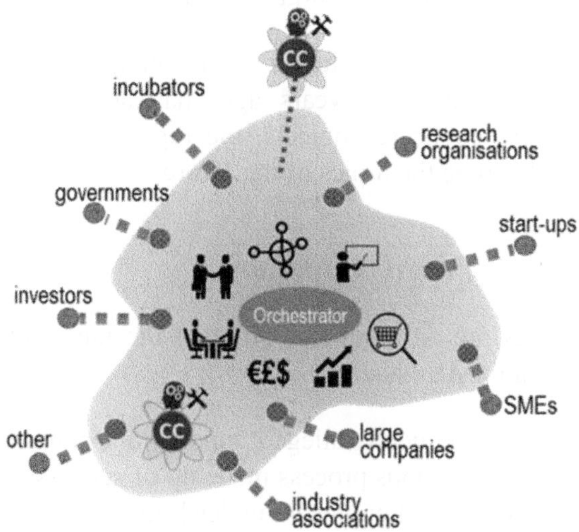

Figure 8.6. *Presentation modes of DIH (source: European Commission, 2017). For a color version of the figure, see www.iste.co.uk/salgues/society.zip*

DEFINITION. A Digital Innovation Hub is the consolidation of various actors within a consortium, associated with a skills center whose goal is to accelerate the implementation of new products and services via the creation of an innovation ecosystem. Innovation is implemented by an orchestrator.

Thus, DIHs use skill centers as tools and are based on ideas of coordination and cooperation, fundamental characteristics of society 5.0.

The European Commission introduced DIHs by means of Figure 8.6.

8.9. The evolution of former innovation organizations

In order to respond to this development, those centers that are in charge of innovation should be adapted in view of achieving excellence and of supporting young firms in their development. The world of "co" implies encouraging networking, resorting to experts and to those in possession of knowledge, having access to training and sharing experiences. All these approaches become more important than support, which until recently had been the dominant activity of these organizations.

A second element in its evolution is the change of topic: ICT, the spearhead of these centers 30 years ago, has accumulated experience. Bioproducts, energy, connected objects, the digital, logistics management and robotics have become the topics of the future.

8.10. Innovation in human resources

Human resources are an important point of the innovative organization and should be recognized as such.

Contrary to what is frequently alleged, innovation should not be managed as a project, but is a continuous process made up of successive projects. That is why hiring researchers and engineers under limited duration contracts is a mistake and a source of failure. Job security is a necessity, but associated with recruitment and selective engagement. Team management is more effective if management is decentralized and the teams are self-governing, but with targets, goals or visions to be turned into achievements. Salaries should be comparatively higher than in other functions and rather associated with organizational performance. The members of these teams usually have high education levels or a broad educational background. In short, in these teams, flat organizations with few differences in status are shown as the most efficient ones.

Co-working is one of the fundamental elements of "co" society and of society 5.0.

DEFINITION. Co-working is a type of work organization which brings together two concepts: a shared workplace and a network of organized workers, in view of facilitating exchange and openness.

Co-working spaces are sometimes called "shared offices". Thus, Fab Labs and living labs are examples of co-working spaces. This is an example of collaborative economics. These co-working spaces become a context which, by fostering exchanges, facilitates implementations that behave as innovation triggers, thanks to the exchanges between collaborators.

These areas are significantly expanding, and many private actors are becoming engaged in offers in this direction. Room rentals have most often transformed into co-working spaces. Office rental agents have felt the importance and the interest of cutting out on urban rentals which had become too expensive. In fact, renting several offices in a flat is more profitable than renting a flat, which is itself more profitable than renting a set. On the other hand, structuring networks and facilitating direct cooperative exchanges are more difficult to measure.

9

"Co" Society

"Co" society is one of the elements that give shape to society 5.0 and to new forms of industrialization.

9.1. "Co" society

"Co" society, where co-construction, cooperation and coordination dominate, is a characteristic stage of society 5.0. Many economic actors have become aware of "co" society, which used to be limited to artists such as Fred Forest in the 1960s.

While "co" man is replacing all previous kinds of humanity, society 5.0 is reuniting all of the features of previous societies.

9.2. The evolution from prosthetic man to the current man

Man is a singular being, given the fact that he thinks and he is conscious[1]. Industrial society manufactured a prosthetic man. Anthropologist André Leroi Gourhan argued that the tool is an extension of the hand. The society of information, knowledge and entertainment that we currently live in is questioning this assertion.

1 Bertrand Vergely, *op. cit.*, p. 33.

Types of society	Types of man
Society 1.0	Prosthetic man
Society 2.0	Civilized man
Society 3.0	Rational man
Society 4.0	Information society man
Society 5.0	Augmented man, improved man

Table 9.1. *Types of man and types of society*

9.2.1. *Types of bored men*

In 1881, Édouard Pailleron distinguished two types of men, before adding a third type: “There are only two kinds of people in the world, those who do not get bored and who are nothing, and those who get bored and are everything. Then, there are those who know how to bore the others.” Psychology places man in the anal stage. The anthropological vision, which we find in Liaudet, Gras or Augé, has favored progress in the classification of men in their reactions to boredom.

9.2.2. *Prosthetic man*

The objects of the industrial world are prostheses, which help man to be more effective. It is placed in a man–technique–object–environment continuum, which emphasizes the couple technique–object. In this case, technique and boredom are linked and they even seem to reinforce each other, which would explain the constant need for innovation, for a better technique in order to become even more efficient.

Man is a being equipped with many prostheses. Let us consider some examples. The car is the natural extension of our limited body, in terms of natural motion. It helps us go further and faster. It is hard to imagine a world without cars. The case of the car is interesting for showing that strange links are woven between technique, prosthesis and death. For men, any technique becomes dangerous; it imposes a search for security, the construction of trust. It makes basic human mortality visible, a mortality that can be statistically treated. That is probably why we seek to cover up those prostheses which are nearest mankind, such as hearing aids or articulated members. Currently, a reverse phenomenon is appearing. Those with a prosthesis show it off with an air of superiority. To be convinced, it suffices to watch the developments of Nicolas Huchet (Bionico project) or of Christophe Debard (MHK Association and the "print my leg" project).

Lars Fr H. Svendsen stated that "anthropocentrism engendered boredom, and when technocentrism removed man from center stage, the problem only worsened."[2]. For him, this could be because technique induces dematerialization of the world, in which things disappear in their pure functionality. Information society has intensified this phenomenon, since machines tend to become outdated before potential users have had enough time to learn how to use them.

9.2.3. *Civilized man*

Becoming a man means becoming civilized, that is, going out from nature. Man should live a dignified sociability with his fellow men[3]. He should become involved in the construction of a world ruled by humanist values. Civilized man as such is opposed to natural man. Hobbes, Levi Strauss and Freud considered man in view of the repression of nature. This led to the birth of the symbolic order. According to André Leroi-Gourhan, humankind, and therefore humans, emerged when man made the transition from the four legs to standing erect. By means of this evolution, man freed his hand and gesture, which genuinely defines humanity. The gesture goes far beyond technique. The same goes for the word, which certainly defines humanity.

9.2.4. *Rational man*

Rational man is inspired by an economic approach. Rational thought makes a distinction between subject and object. On the contrary, mythical thought merges these two concepts into one. In order to attain happiness, the rational man indefinitely accumulates material goods and values, pursuing a logic of self-interest. As Ladwein pointed out, economy has contributed two ideas to marketing. First of all, choices are rational and motivated, because they are the object of deliberations among competing offers. We can then measure choice behavior from the characteristics and functions of the

2 Lars Fr. H. Svendsen, op. cit., p. 123.

3 Bertrand Vergely, *op. cit.* p. 189.

competing offer[4]. As observed by Marc Augé, the individual is free to consume what he wants, but his choice is limited by the apparently distinct range of products that suppliers offer him[5].

Rational man is an imperfect answer. Any form of reasoning confronts life, is a crash and is a power grab, which results from man's anxiety of not possessing goods. Marc Augé noted that the individual is not really free to consume. Advertising, the various forms of credit, fragility and renewal of products limit his freedom of choice.

9.2.5. *Information society man*

The man from the information society uses special robots. Computers reason as humans do. Artificial intelligence helps man to choose between different products on the market. This is the role of websites, which make it possible to know the availability and price of products.

9.2.6. *Augmented or improved man*

This concept refers to technologies seeking to increase human capabilities beyond the scope of their natural biological evolution.

In this context, we can distinguish different approaches that seek to repair or transform augment capabilities. Another dichotomy is the augmentation of man in his physical capabilities versus an increase in his intellectual ones. Society 5.0 is involved as an actor in both cases.

9.3. The split between boredom and innovation

The problem of boredom has no solution, which precisely makes it a real problem, and hence it facilitates the lives of marketers. In contrast, innovation is always a solution to a problem. Boredom is a boundary for utopia. Utopia cannot thoroughly happen. If utopia is achieved, then boredom is created, which finally destroys any utopia from the inside. This shows the importance of attitudes like those proposed by Scharmer in U procedure.

4 Richard Ladwein, *Le Comportement du consommateur et de l'acheteur*, *op. cit.*, p. 24.

5 Marc Augé, *Où est passé l'avenir ?*, *op. cit.*, p. 9.

9.4. New innovative strategies

"Co" society is leading to a redesign of innovative strategies concerning several points.

9.4.1. *Innovation must be everywhere*

Within "co" society, innovation is present everywhere.

Within "co" society, innovation is a topic covered in several publications. The Center for Young Leaders deals with innovation on a daily basis. For them, the key to the success of firms is to make a difference. For this reason, innovation is necessary, but its life span is short (barely a few months) because of competition[6]. François-Marie Pons and Marjolaine de Ramecourt have praised the principles of innovation at all levels of a firm. This topic has also been treated by Isaac Gertz and Allan Robinson, or Bob Schmetterer. For them, it is the firm that generates innovation. For this, it is necessary for the firm to be innovative and to encourage each employee to behave according to this orientation. Participatory innovation is desired, and the existence of "unconventional bosses" might be an indicator of success. Encouraging innovation is vital. The problem of these innovation theses is that they are poorly interested or not interested at all in the consumer.

9.4.2. *The end of the dynamics of jealous marketing*

Jealousy is the desire to possess what the other person has[7]. Marketing exploits jealousy. In fact, by means of desire, it transforms exchanges into something dynamic and, therefore, energizes production. When we have desires, we "want to possess", we most often desire what the market offers and the desiring subject who wishes to access this economic life does so by becoming one of its actors. Sometimes, this phenomenon turns out to be perverse. Sometimes it happens that we experience desire only to make another person jealous and that he, in turn, also experiences desire, so as to make a third one jealous. Frustration appears, as opposed to jealousy. The fable by La Fontaine, "The Fox and the Grapes", is full of teachings. Since it cannot reach the grapes it wants, the fox begins to denigrate them, to kill what it desires.

---

6 CJD2003.

7 Bertrand Vergely, *op. cit.*, p. 107.

9.4.3. *"Co" society as a means for understanding the consumer*

Finally, to limit the understanding of consumer behavior just by analyzing his satisfaction level is certainly reductive. Besides, it cannot account for the methods intuitively discovered by product managers, who are constantly varying the shapes, as well as the composition and packaging of products. Responding to boredom is probably one of the explanations.

9.5. Porter's strategic model

Michael Porter's strategic diagram seeks to describe a strategic model.

9.5.1. *The notion of strategy and of strategic model*

DEFINITION. Strategy is the way the firm chooses its battlegrounds and invests its resources in them. This helps it to maintain or improve its competitive position, based on current or future changes in the environment.

Porter's strategic model could inspire the analysis of a possible strategy, as shown in Figure 9.1.

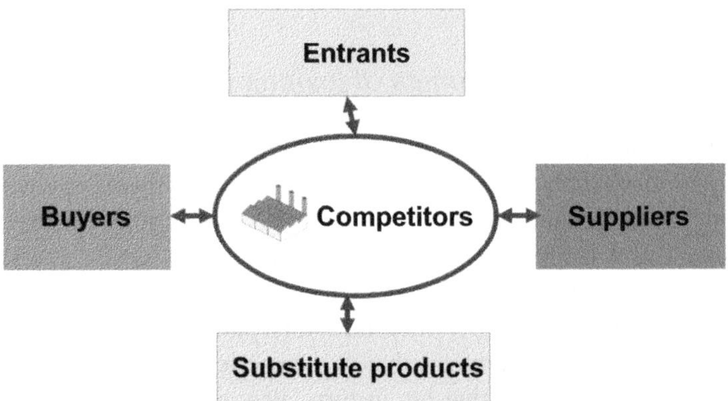

Figure 9.1. *Porter's model. For a color version of the figure, see www.iste.co.uk/salgues/society.zip*

The analysis of sectors as proposed by Porter represents the five forces of competition. The example of telecommunications is quite simple: competition takes place between operators. The new entrants are cable television suppliers and those actors who transmit data. Suppliers wish to provide services. The customer seeks price reduction, whereas substitute products are essentially related to transport.

9.5.2. *The concept of value chain*

We can reuse Porter's analysis to examine firms, with the help of his concept of value chain.

DEFINITION. The value chain refers to activities that create value (main and support activities) and the margin.

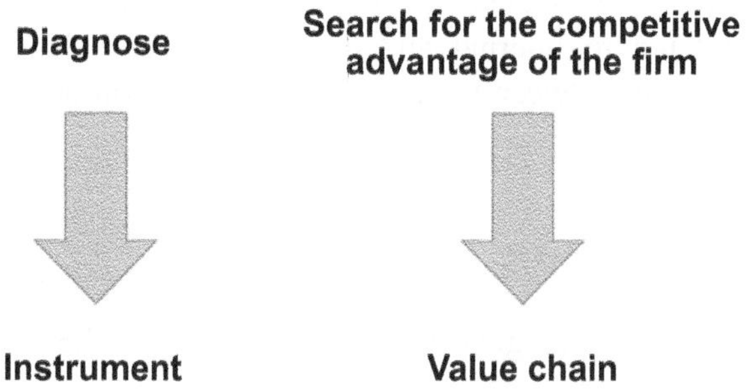

Figure 9.2. *Example of the transition from diagnose to the instrument*

The comparison of the firm's value chain as regards:

– competitors, makes it possible to take the competitive field into account;

– the provider, can help identify interdependencies and

– distribution channels, makes it possible to establish the competitive advantage.

According to Porter's terminology, all the activities of the firm and their interactions constitute the firm's chain of value[8]. There are nine types of

8 Michael Porter, *L'Avantage concurrentiel. Comment devancer ses concurrents et maintenir son avance*, p. 49, InterÉditions, Paris, 1986.

activities that can create value, which are classified into the following two groups:

– the main activities, grouped into five categories: internal logistics, production, external logistics, marketing and sales;

– support activities, including four categories: the firm's infrastructure, human resources management, research and technological development and supply.

As observed by Porter, competitive advantage often derives from the many connections between activities, as from the activities themselves. In other words, the interdependence of activities is a key factor of success, which should be coordinated.

To these valuable elements, we should add quantitative coordination, which is related to a system of cost control, as well as qualitative coordination, related to internal and external power balances, which are essential for choosing between total or partial internalization and cooperation agreements.

9.5.3. *Porter's three basic strategies*

Porter's three basic strategies are described in Figure 9.3.

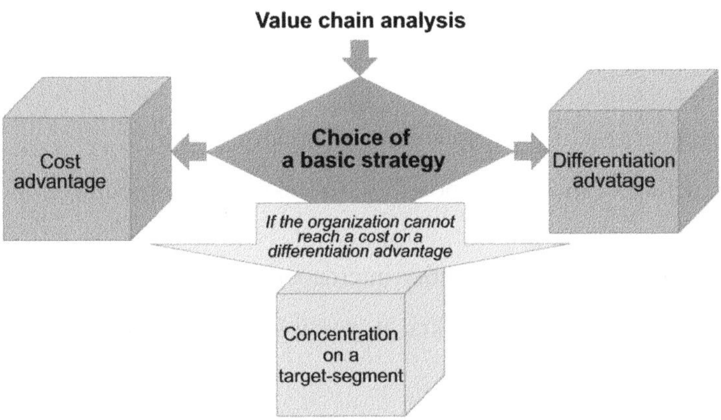

Figure 9.3. *Porter's strategic diagram. For a color version of the figure, see www.iste.co.uk/salgues/society.zip*

9.5.4. *Cost strategic advantage*

The cost advantage is one of the three types of competitive advantages that a firm may exert. It is important, however, to combine it with a differentiation strategy, since the firm needs to keep its costs close to those of competitors. Cost analysis is related to activities that create value (not to the firm's customers). A firm obtains cost advantage if it succeeds in carrying out its value-generating activities at a lower cumulative cost than its competitors. This could account for the success of the so-called low-cost services.

In order to analyze costs, we should isolate those activities:

– whose operating costs represent an important fraction of the firm's costs;

– whose costs involve different evolving factors (e.g. the promotion cost is different from the advertising cost, which is subject to scale) and

– towards which competitors behave differently.

This approach has led to the development of management methods such as lean management.

9.5.5. *Differentiation advantage*

Differentiation advantage is the second basic strategy: a firm is different from its competitors when it manages to acquire a unique characteristic, to which customers attribute a certain value. Differentiation is generally expensive. Besides, it should not be limited to the physical product, since it is necessary to reach the global product. In the past, differentiation used to be aimed at design and branding. "Servitization" is one of the new possible sources of differentiation that should be oriented toward the satisfaction of customer needs. For this, the firm should consider the value chains of customers.

Differentiation can take many forms. Differentiation among industrial, commercial or institutional customers reflects their strategy and the steps to carry it out. Differentiation among households reflects the attitudes and needs of their members, as well as how to reduce customer costs and improve customer performance. For this, it is necessary to understand what the customer desires: understand how the customer perceives value, that is, the criteria used by customers in order to calculate the value created by the firm (advertising, packaging, fame, reputation). Finally, it is necessary to know the purchasing criteria of customers, which can be of two types:

– use criteria (product quality, delivery deadline, technical assistance in connection to the real product, the product's prestige, etc.), which mainly aims at usefulness;

– signaling criteria (advertising, packaging, brand image, etc.), which mainly aims at desirability.

9.5.6. *Focus strategy*

Focus strategy is the third basic strategy. If the firm is unable to gain an advantage by costs or differentiation across the sector, it focuses on a target segment. It is therefore necessary to previously study segmentation; in fact, it is the central point in the choice of focus strategies.

Segmentation basically aims at the delimitation of a sector (a market selling the same or neighboring products). The choice of a strategy requires segmentation, because products in which customers perceive distinctive characteristics have a strong impact on how the firm can gain competitive advantage. The differences between product varieties or customer differences act as factors of segmentation if:

– they modify one of the five forces of the competition;

– they affect the firm's value chain.

In practice, we group product and customer varieties into categories that reflect the most salient differences. We can appreciate the following four segmentation variables that help us define the segments:

– the different varieties of products that are produced or could be manufactured;

– the different types of customers who buy or could buy;

– the different distribution channels used by customers;

– the geographic placement of the customer.

The viability of a concentration strategy depends on the strength of copycats, the possibility of having access to substitute products and the existence of large-target competitors.

9.5.7. *Development pathways*

There are two important pathways to development. The first one is deduced from the analysis of the pair product–market and is called "proximity development". The second one is based on takeovers and is called "consolidation development".

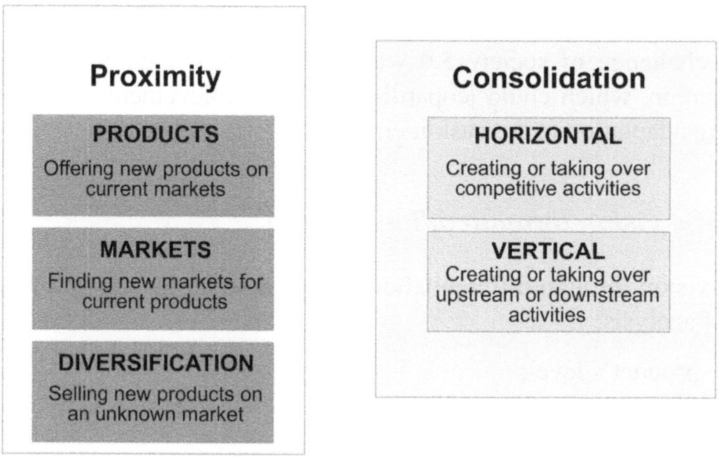

Figure 9.4. *Pathways to strategic development. For a color version of the figure, see www.iste.co.uk/salgues/society.zip*

9.5.8. *The origins of market massification*

The massification of markets is a fact. It is necessary to understand the origin of this phenomenon. It aims at reaching economies of scale. For this, it is necessary to combine a concomitant decline in prices and an increase in production.

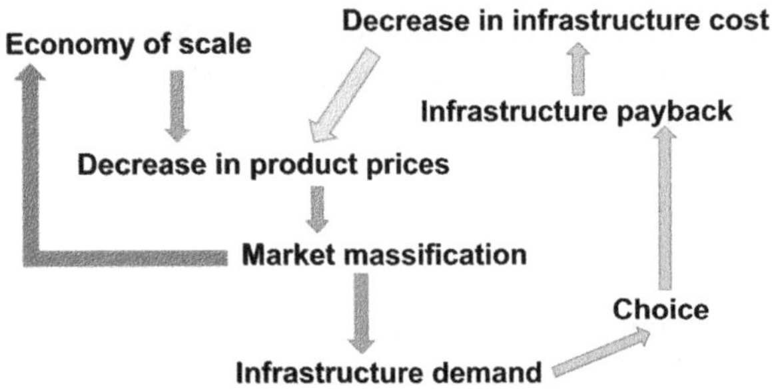

Figure 9.5. *Mechanism of massification. For a color version of the figure, see www.iste.co.uk/salgues/society.zip*

The challenge of society 5.0 will be to avoid a natural and automatic massification, which could jeopardize the key achievements of society 5.0, including adaptation to the customer.

9.5.9. *The vision through differentiation*

The vision through differentiation can be defined by the following three levels of analysis:

– the product's level;

– the service's level;

– the firm's level.

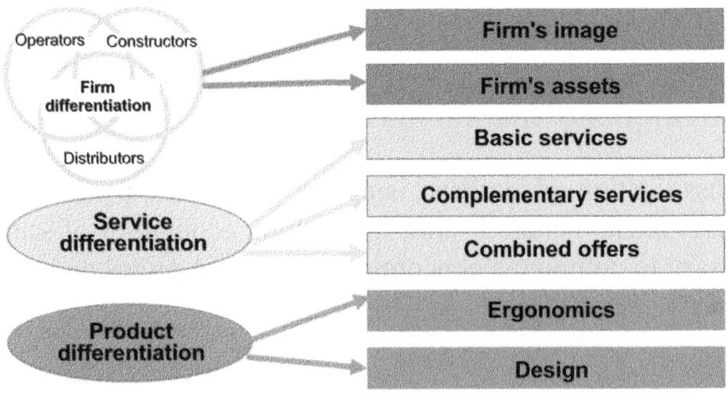

Figure 9.6. *Differentiation levels. For a color version of the figure, see www.iste.co.uk/salgues/society.zip*

9.6. Useful partnerships

Traditionally, massification of markets has involved the creation of partnerships or company takeovers. For manufacturers, the emergence of a mass market provokes certain difficulties that can be faced by means of standardized solutions.

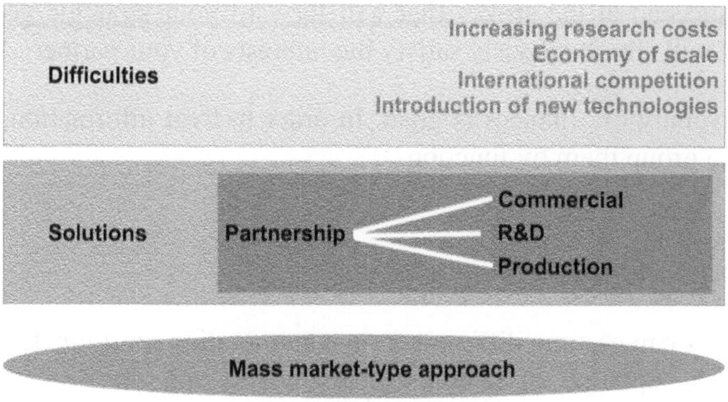

Figure 9.7. *The mass market type of approach. For a color version of the figure, see www.iste.co.uk/salgues/society.zip*

Partnership is imposed by the mass market. It is essential:

– to reduce and share development costs;

– to save time;

– to reduce human resources to the essential minimum;

– for sharing technology;

– to increase the commercial presence of every actor involved.

This phenomenon is all the more important in order to bring together skills from different fields and implement such a network. For example, it is possible to increase the commercial presence thanks to franchisees, that is, partnerships with other vendors. This type of action is a characteristic of "co" society.

9.7. Different types of alliances

In the broad sense, according to the typology of Charles Wiseman, the following three specific forms of strategic alliance should be considered: acquisitions, mergers and agreements[9].

> "It takes the skill of a diplomat and the cunning of a guerrillero, to succeed at the strikes that will forward your strategic goals and will simultaneously satisfy the interests of your partner..."

The diversity of alliances is great. In order to treat information, it seems essential to group them by function.

9.7.1. *The conditions of alliances*

Alliances can assume different forms: cooperation, participation or even voluntary acceptance of shared agreements. Long considered only in their juridical aspect, these practices nowadays have a strategic dimension. In the case of strategies based on technology, particularly in our case, close links must be maintained with multinational and university research labs.

In an industry where technology plays a decisive role and where, by the award of public contracts, the State is in a position to act as a market referee, the success or failure of firms is by far more closely linked to the State's interventions and its ability to generate better profit in their favor than what they owe to the competitive strategies they are likely to implement. Different authors, including Dussauge and Ramanantsoa, have published typologies of strategic alliances[10].

---

9 C. Wiseman excluded mergers and acquisitions known as "diversifications", since they go beyond the scope of his analysis.

10 Pierre Dussauge and Bernard Ramanantsoa, *Technologie et stratégie d'entreprise*, pp. 149–154, McGraw-Hill, New York, 1987.

Currency trading technology	Alliance's goal technology	
	Alliance is limited to R&D	Alliance also concerns industrial activity
Synergy alliance Distribution agreements Exchange of patents	*Pre-competitive alliance* Common laboratories	*Industrial cooperation* European projects

Table 9.2. *Typology of alliances based on Dussauge and Ramanantsoa*

Vertical or horizontal integration strategies have reached their limits in many industries and hence a specific phrase has been repeated in unison by many business leaders: "We have entered the era of alliances."

According to Guy Crespy, from the observatory on industrial strategies, alliances during mergers/acquisitions reflect the difficulty of firms to finance their own developments, where growth is uncertain, and the barriers to entry are high. This is the major axis in the information industry firm strategy. Motivations are diverse.

9.7.2. *Strategic alliance through fusion*

The degree of equity as well as the lifetime of alliances varies greatly. In the automobile sector, GM and Toyota are involved in equal parts. In the telecom sector, France Telecom and Deutsche Bundespost, respectively, handle 40% and 60% of Czech telecommunications. As we can see, proportions may be very diverse and involve two or more participants. However, it is also necessary to observe that these alliances tend to loosen over time.

9.7.3. *Strategic alliances involved via the execution of an agreement*

For example, firm A authorizes firm B to distribute the product that firm A manufactures. Similarly, firm A manufactures firm B's product under license and firm B sells it. These are two very common cases and there are a multitude of them, as we will see later. This presentation will only refer to strategic alliances, regardless of total mergers and buyouts. For synthetic reasons, it is nonetheless possible to group agreements that show similarities of form.

Research was made on a database where all the agreements passed in 1960 are listed, including the global groups exclusively belonging to the small triad. The sectors affected in this panel are telecommunications, informatics, electronics, audiovisual, software and semiconductors. The guiding principle had been to group agreements passed from 1987 to 1991 in five sectors. This task was performed arbitrarily, but in accordance with the rules established by the industry; we have been able to establish the following convention:

– research and development agreement;

– production agreement;

– distribution agreement;

– licensing agreement;

– cooperation agreement.

All alliances are classified under these five sectors, which represents a total of 1,073 agreements over the reference period. Taking these elements as departure point, it is possible to proceed to a data analysis, which will hopefully enable us to discover the significant connections between certain industrial groups. In light of this analysis, it is possible to establish a typology of alliances between the different actors involved, which will be illustrated by examples afterward. Let us objectively point out that as much as we have the agreements, we do not have the eventual split-ups, which might soften the rigor of the analysis.

9.7.4. *Alliances through the integration of products*

The first type includes simple turnkey alliances. This is the case of Computervision Corp., leader in CAD software, which signed an agreement with IBM. The latter provides its software, thus developing its sales, and the latter offers a computer with added value. Geisco, belonging to General Electric, also signed an agreement with IBM. It offers a provision allowing IBM to sell a PC with added value and to later sell this business to a Chinese firm, which endorses the brand.

Turnkey alliances are more complex. They involve several players. For example, John Hancock Mutual Life Insurance Company provides an integrated service to its agents. Thanks to agreements signed with IBM, Lotus used to offer a software-integrated solution at a preferential price. Still, this alliance did not succeed in outwitting the challenging Microsoft.

Alliances may also concern the distribution of products. They make it possible to multiply distribution channels. Electronic Data Systems (EDS) provide teleprocessing and installation management services. Alliances with Hogan Systems and Cullinet software authorized the extension of the range of systems offered to customers. Here again, firms penetrated market segments that had previously remained closed to them.

9.7.5. *Determinants of an alliance*

Firms in the industry become allies for different reasons, and we can group them together as follows:

– research and development agreements;

– production agreements;

– distribution agreements;

– licensing agreements;

– cooperation agreements.

The results of an alliance are linked to the definition of strategic fields of activity. This approach particularly involved Michel Godet (1991) and the Strategor multiauthor book (1988). From it followed a kind of tangle of alliances. One of the goals of any firm is to ensure its sustainability. However, at present, this thesis is inconceivable without an external growth strategy. Nevertheless, the cost generated by such decisions no longer allows enterprises from different sectors (particularly the technological one) to choose buyouts, but rather alliances. Only large companies may resort to buyout.

Synergic effects are systematically pursued. Synergy makes it possible to proceed to a joint research effort, to exploit technology developed by a partner or other entity.

9.8. Typology of firms (according to Kotler)

Following this classification of strategic approaches, we cannot avoid the typology that is at the core of the discourse of the majority of specialized articles. There are some similarities to Porter's approach. However, this is a more sporty and direct approach. Philip Kotler[11] founded his typology of market shares and distinguished the following four main strategic directions:

– the leader;

– the challenger;

– the follower;

– the specialist.

9.8.1. *The leader's strategy*

The leader's strategy is followed by the firm with the highest market share (around a double of the "second"), thus exercising a dominant position. In addition, it represents the reference in its sector for customers, suppliers and competitors. The leader is neither necessarily the most innovative firm nor the one that determines market prices.

Its strategic procedure involves:

– developing global demand;

– locking the market;

– taking profit from the effect of experience.

For example, IBM is still the world leader of computer manufacturers. This is the reference on the market. However, Sun Microsystems, the leader of the manufacturers of high-end workstations, has now disappeared, while actors are becoming large in a market that has been abandoned by IBM, which is that of PCs and tablets.

11 Philip Kotler and Bernard Dubois, *Marketing Management*, Publi-Union, Paris, 1985.

9.8.2. *The challenger's strategy*

The challenger's strategy is the strategy of the "second ones" who have the ambition to acquire a dominant position in term and become the leader. The questions that the challenger faces in order to attain his goals lie at the levels of:

– the choice of strategic field of activity that should be tackled;

– the means to be used;

– the assessment of the leader's capacity to respond.

In this way, the firm Fujitsu moved upward, after buying ICL in July 1990 and attained the second position, before DEC, NEC and Hitachi. It currently challenges IBM. However, it will have to detach itself from the activities that bind it to IBM in order to find its technological independence and thus face Big Blue on equal terms. This strategy has not reached its goal[12]. For example, operating systems based on Microsoft architecture have been challenged by the arrival of Google's Android.

9.8.3. *The follower's strategy*

The main objective of the follower is to profit from the situation. The follower profits from the experience of the leader and often implements lower prices. The strategic actions implemented mainly focus on R&D (buying licenses, OEM agreements, etc.) or on the customer (fidelization, etc.).

Thus, the strategic direction groups the majority of the firms in the computer sector offering IBM-compatible materials: these are, in the specialized jargon, clones. More recently, we have seen the emergence of smartphone suppliers, which are "Android-compatible".

12 M. Cross, "Fujitsu, je t'IBM moi non plus", *Challenges*, October 1991.

9.8.4. *The specialist's strategy*

A firm that has adopted a specialist strategy, called "strategic niche", bases its approach on a very sharp market segment. According to Kotler, the main difficulty for this orientation lies in the need to find a match between the distinctive competence of the firm and the specific characteristics of this market segment.

As shown by Le Duff and Maïsseu in their reflection on the evolution of metatechnology, the denser the technological mesh, the more important and powerful will be the synergy progress generating effects. Thus, the development of information technologies makes it possible to witness the birth of industrial projects on restrained and innovative niches. The notion of metatechnology refers to bisociation theory, whose method it sometimes applies. In other cases, metatechnology is born through alliances.

As we have seen on the previous typology, workstations employ a sharp knowhow, which encourages the birth of newcomers. In this case, it was the development of networks and graphic interfaces, among others, which gave rise to a large-scale industrial project. However, electronic components are also extremely popular nowadays. Let us recall that this is now achievable thanks to a continuous process of component integration and miniaturization and, as we previously mentioned, the development of architectures.

10

The Challenges of Localization, the Market, Skills and Knowledge

The technologies of society 5.0 are modifying the current society in relation to the following two localization phenomena: (1) a situation in which localization is losing its importance; and (2) actions in which localization has become crucial. In fact, localization is related to the issue of available (human) skills, as well as available knowledge.

10.1. Localization is increasingly losing its interest

Every social revolution is characterized by a change in social relationships, which is a consequence of technical innovation, and this is not an anecdotal phenomenon. The industrial revolution, which accompanied society 4.0, entailed:

– a mechanization of the individual, an example of which is the possibility of chain work, which led to a repetition of the same task all day long, and

– a strong cost reduction thanks to an increase in the volume of production and an improvement in efficiency.

The consequence that ensued in society 4.0 was the need to produce a certain amount so as to be competitive, and wrongly, this became a synonym of "industrialization". Because of its capabilities, society 5.0 seeks to detach itself from this need for volume.

A second type of consequence is the search for large amounts of elicited energy so as to intensify the production process. Thus, the proximity of coalfields was one of the assets to boost industrialization. The existence of large population basins responded to the need for labor force, which helped to avoid resorting to foreign workers, which would have provoked societal consequences. In his book *La Grande Divergence*, Kenneth Pomeranz explained the differentiated trajectories between Britain and China at the end of the 18th Century, despite the fact that they were roughly at the same level, technologically and financially. The key element is localization, an important point of society 4.0.

Electricity, robotization and information technology have made the choice of location truly obsolete. One of the consequences is that the challenge for society 5.0 becomes the ability to locate organizations within itself.

Robotization affects all economic sectors; its introduction threatens both low-skilled jobs and industries in countries with low production costs, either in terms of labor, the access to resources, or due to special regulation. The example of the textile sector is particularly striking. The labor market became highly digitized. Upstream, software makes it possible to determine the model's patterns and to obtain 3D images.

As a consequence, the remaining tasks modify social relations. A reflection on these changes in society 5.0 could provide raw matter for several books.

10.2. New practices related to the lack of importance of localization

In the book by Allard, Creton and Odin dedicated to the mobile phone, the topic of the end of certain modes of communication is raised. For example, direct communication with the desired person and the knowledge of his or her location entailed the disappearance of phrases such as “I will put you through”, a sentence which used to be pronounced when a call was placed on the family phone.

10.3. The importance of reconstruction

When a firm has to rebuild itself in industry 4.0, it is often obliged to abandon a historical site, which no longer corresponds to the standards and capabilities related to the construction of the factory of the future, whose architecture must match some special requirements.

For example, the firm Latécoère, which is planning to invest 30 million Euros in the factory of the future, is forced to give up its historic seat of Périole Street in Toulouse. It is moving to a suitable field in Montredon. The size of a production unit of this type generally requires a single block of area 5,000–10,000 m^2, which cannot exist in historical sites.

10.4. Changes in market shares: why and how?

The logical path is to expand the market share. There are four reasons for this. The first one is the law of experience. With an increase in production, the total cost per unit of a product decreases. This situation is related to experience and the learning phenomenon. The second reason is related to the fact of the firm's interest in reaching a certain level of production. The third reason places the firm in a strong position in front of its customers and suppliers. Finally, the increase in the market share increases product visibility and, consequently, brand recognition.

For this, there are several solutions, including extending the range of products, improving the product thanks to new techniques or quality, innovating the product, increasing commercial spending, bringing coherence to the "mix" or segmenting and differentiating products more finely.

Therefore, it is often necessary to defend our own market share. For this, there are generally three options: the first one is to attack, by improving the elements of the "mix". The second one aims to fortify oneself, by saturating the market with different brands. This solution entails risks; in fact, it has come to be known as "penguinization". It comes from the image of a penguin army: the consumer realizes that the product is always the same. Finally, the last solution is to choose open confrontation either through a price war or by strengthening commercial actions.

10.5. The issue of skills and knowledge

The issue of skills and knowledge is complex, but refers to the handling of intellectual capital and to the role of the couple training-information and the necessary changes that should take place in management, including the commitment from the actors, changes in standards and programming and planning modes. There is no modernity without knowledge; there will be no society 5.0 without new knowledge. The tool has always existed and adapted to the conditions of knowledge that each historical moment has made possible. And a computer is as dumb as an abacus without a human giving meaning to it. To transform public administration or firm by digitizing and dematerializing, conventional processes will not make rules "intelligent" by miracle. Unfortunately, many States and business leaders have taken this path. On the contrary, technique can be used for creating new solutions as well as new rules to deal with complex problems. For this, knowledge should be used in order in view of modifying skills.

10.6. The notion of intellectual capital

This notion was formalized in a very popular article by Mohan Subramaniam and Mark A. Youndt, "The Influence of Intellectual Capital on the Types of Innovative Capabilities"[1].

DEFINITION. Intellectual capital is made up of three elements: human capital, social capital and organizational or tactical capital.

DEFINITION. Human capital is based on creativity, education, expertise, intelligence and talent.

This human capital may be improved thanks to ongoing education and information, which we will analyze later.

DEFINITION. Tactical capital is represented by patents, licenses, databases, procedures, organizational culture and systems, particularly, information systems.

---

1 Mohan Subramaniam and Mark A. Youndt, "The Influence of Intellectual Capital on the Types of Innovative Capabilities", *Academy of Management Journal*, 48:3, p. 450–463, June 1st 2005. Available at: http://amj.aom.org/content/48/3/450.abstract.

This tactical capital may be improved thanks to the implementation of changes in the tactical section of the organization, particularly in the marketing sector. We will devote special chapters to this question later.

DEFINITION. Social capital is the ability to cooperate, share information, learn from each other, interact and exchange, create alliances and use knowledge from unusual and diverse sources.

In a traditional and certainly rather simplistic manner, three outlooks on social capital are considered.

The tools vision considers implementing a social network or a network organization as we have described. Social network software should suffice.

The sociological vision is interested in the totality of strong bonds that hold a community tight. It also studies attributes such as shared standards, trust and reciprocity. This vision seeks to remove everything that:

– isolates or blocks information;

– hinders the understanding of external challenges;

– makes people complacent;

– encourages conformism;

– keeps the firm away from innovation.

The managerial vision seeks all the skills related to cooperation or mutual or common learning. Then, the organization should make it possible to encourage the sharing of information among work teams. This is necessary to favor the emergence of innovation and that which makes organization adhere to the principles of "co" society. It is also necessary to identify strong and feeble links in networks. Finally, this approach should allow everyone to contribute with new ideas.

Intellectual capital helps to characterize radical and incremental innovations, as well as frugal and social innovations.

Type of innovation	Features
Radical	Industry dynamism High expenditure in R&D Strong social capital Strong human capital High interaction between social capital and human capital
Incremental	Industry complexity Important social capital Operational tactical capital
Social	Importance of social and human capital Strong interaction with future customers
Frugal	Actual reduction of costs Interaction with future users in order to understand necessary functionalities

Table 10.1. *Innovation methods and intellectual capital*

10.7. Changes in operational marketing

Operational marketing has changed a lot due to the Internet. Not only dematerialization, but also the fact of being able to bring forward something believably human is important. The little virtual characters that guide us on the Web are the ultimate achievement of this.

The Internet has become a source of promotion and advertising to obtain data about the consumer in return. This is particularly important in areas where commercialization platforms become must-sees for purchases, as in the cases of leisure and the hospitality industry. Thus, in Best Western, the budget devoted to Web media represented 0% of expenditure in 2000, 5% in 2003, more than 50% in 2009 and 80% in 2014. The evolution in the behavior of managers had never been this fast.

Societies now measure Internet marketing. In fact, there are new firms devoted to marketing studies. The goal is to attribute marketing signification to data obtained on social networks. There is a plethora of data, but these arouse utility problems. The best way to become competent in this field is to "do things". For the marketer of society 5.0, it is necessary to have the basics that university provides, but logic is not transmitted within the classroom, it must be discovered. Curiosity and the tools for making sense are certainly new skills. To paraphrase a director of a company in this field, "firms in society 5.0 have to go even beyond the Internet!"[2]

In old firms, it was the manager who handled marketing campaigns. At present, it is necessary to handle data in the long term, the aim being to integrate all into managerial processes and not only in the marketing sphere. New trades are renewals of previously existing businesses. The managers of a bank agency can run their own blog, which changes the job.

In some areas, such as high-tech, the specialized press has disappeared (is on vacation), and the Internet becomes the only solution to become informed. New professions are born, such as those interested in groups of customers or prospects, often called "community managers".

10.8. Intrusive marketing

The intrusion of brands in private life has become a reality, but appears as a possible limitation to the interactive and customized communication of the future. Commercial websites introduce scoring systems, in which customers can rank brands or suppliers, as in online selling platforms such as eBay or Amazon.

The intrusion phenomenon goes far beyond the construction of an increasingly interactive relationship with brands. Nowadays, intrusion has gone as far as implementing car devices, which analyze the risks associated with each driver. The insurance bill is a function of motor behavior, the "pay as you drive", which is actually banned in many countries such as France. A Spanish bank allows the trader to send a text message during purchase thanks to special offers such as *bankinter* and *extrafuturecard*, a kind of fidelity card that follows the behavior of consumers at a shop rejected by the consumer. There exist other forms of the so-called incoming intrusion: text messages, information about selling point, mailing, etc. Outgoing intrusions are more recent: tracking, Internet analysis via smartphone information. Coincidental intrusions are, for example, the offers of complementary goods after the purchase of a product. In fact, resistance against intrusion is only beginning to become organized. Awards are given to firms that respect personal data the least. Among data, there is shareable information (age,

2 This sentence belongs to Laurent Flores, founder of CRM Metrix, now MetrixLab, subsidiary of Macromill and present in more than 90 countries.

gender, address, etc.), taboo information (salary, images, friends, etc.). The interest of social network sites such as Facebook is that they can temp the consumer to reveal taboo information that he/she did not want to share at the beginning.

Another way of obtaining consumers is to create "incentives". This refers to getting information in exchange for advantages in terms of purchase power. For this, the firm needs to be as transparent as possible. To do this, the firm must appear as legitimate; invite win–win situations, perform more focused operations and avoid the impression of saturation that the consumer may have.

10.9. The use of acquired knowledge

Jacques Perriault[3] wonders about the use of compendiums of good practices and application profiles. First, we should consider that these devices appear as all-comprising instances, which ensure that one has a common cycle of knowledge. Then, good practices serve as point of departure for debates, which are uses and normative practices.

For F. Cooren[4], these documents were just instructions, procedures and action lists to be carried out. We follow directions, and those who created them are carrying us somewhere. For the one using them, there is a "presence–absence" effect. The one who shapes the material is an "auctor", a term which shares the same Latin roots with author and authority. Therefore, this leaves the path open to normalization practices, but invites a reflection on the nature of regulations as disciplinary practice.

Formalized knowledge can be classified according to the following three approaches: demos refer to a bodily discipline and personal development; digests need rewriting and ownership, as well as handling writings and automatisms are inscriptions on the working material itself through special mechanisms, data or programs. Demos, digests and automatisms are certainly not standards.

---

3 This is an idea prompted by Jacques Perriault.

4 F. Cooren, LOG, University of Montréal, introduced this theory at Lille on April, 2nd and 3rd 2009 within the framework of the Org&CO Research days and in his article dating from 2006, which can be consulted in the bibliography.

According to David Charasse, followed by Pierre Delcambre (1997), there are different types of writing practices. The simplest practice is branding, which is often only a symbol of property. The text layout is more advanced, but less strong than formalization, which entails predicting how texts will be written. Standardizing requires formalization. There is a danger in reasoning when talking too fast about text layout. There is another danger: stating, in contrast, that writings are standards. According to Delcambre (1997), socialization means learning "standardized things". In the same spirit, the position stating that know-how is to internalize standards may be operational. Therefore, we should not eliminate the social importance of standards.

10.10. Identification of regulations in documents

Regulations can be made by "additional" writings for products, such as manuals, for example. Some manufacturers produce these writings themselves, which are then displayed by their authors "at the exterior" of the product, on the packaging or the instructions. In other cases, the rules become "registered" on a close environment, such as message boards. Finally, benchmarks are controlled and registered in the environment. It should be noted that most often these concern bans.

A regulation can refer to a more distant written rule, either in time or in an external regulation regarding the organization. This is often present in the form of a text. It is more or less "unfolded" or explicit. When we can access the texts themselves, the regulation is named, and it can be extracted, quoted or duplicated *verbatim*. In this case, human relays are used, they explain the rule and the form is unstable and requires specific implementation.

The rule can be technologically recorded. Thus, the mechanics is more or less binding. In this case, there are perceptual mechanisms encouraging us to follow the rule, as in the example of the car's speedometer.

In these three cases, the major difficulty is the passage from regulation to the norm.

10.11. Identification of forms of commitment

The question is how to identify the forms of commitment in regulations. This question precedes that of regulation distribution, chiefly a political question, especially when the regulation is recorded on the technical object.

First, the forms of commitment can be analyzed within a popularization mode, which brings us back to the dissemination of normalization culture.

The other path is to identify the forms of commitment within actions themselves. The mHealth alliance is encouraging "new innovative thinking", by fostering public/private partnerships. For this, the alliance has defined certain actions, which correspond to specific needs. Each of these meets multisector challenges in view of reducing the effect of pressure currently existing on care systems in different countries. This means mobilizing popular philanthropies so as to raise further awareness on existing illnesses.

10.12. Implementation of normalization

The STANAG is implemented to foster best practices. In France, the TMED network is deployed by CNES in order to diagnose and treat through satellites, thanks to a mobile telemedicine station[5]. Tests have been conducted in Guyana to ensure the reliability of the system. Spain has implemented a TMED system, which enables real-time remote consultations with the specialists of leading Spanish military hospitals[6]. The Polish telementoring system includes a camera in operating systems to help the doctor remotely. The United States has recently developed a joint network of teleradiology, which connects the six military medical centers. In Germany, a powerful telemicrobiology remote system has been developed.

On the contrary, Com communication 689 (2008) from the European Commission has had no real development. Similarly, the Continua forum has been less prolific, and only a few specifications have been produced, particularly for products like oximeters.

5 Available at: http://www.cnes-csg.fr/web/CNES-CSG-fr/3851-la-telemedecine.php.

6 Available at: http://www.lukor.com/not-esp/nacional/0801/29143854.htm.

10.13. Organizational consequences

The second aspect of the industry of knowledge concerns questions generally associated with digitization and its exploitation in terms of risks to individual freedoms, exchange standards and the traceability of transactions. As the regulation of tenders for the CNRS indicates, "the challenge is to not blend many questions into a single approach (questions about trends, strategies carried out by actors, situations in connection with disciplinary traditions and innovations or emerging forms of information and scientific communications)", but to be immersed within a "communication theory". The expected organizational consequences are mainly changes that impact data and machines. The expected organizational consequences are mainly changes that affect data and machines. Indeed, norms and standards govern the use of the networks, exploitation of which is likely to improve or stiffen their uses. They are translated in the form of data standards or in the form of program standards. We will successively examine each of these points.

10.13.1. *The norm as an agent for contextual change*

To change means to move from Fordism to the customer, from government to governance. Government is a situation in which hierarchy and responsibility prevail. Governance is related to the plurality of actors.

According to Didier Chauveau[7], norm is a condition of collective efficiency, related to a current situation due to the shift from government to governance. This shift requires a redefinition of the core business, which is visible in the orientation toward service quality. It is a tax and investment strategy in such a way that activities are regulated. This orientation requires an update of the problem, which is the result of the model's logic model and the logic of the process, which became essential in a context of redeployment of organizations.

The major issue is that we need to rethink the conditions of effectiveness in this new context, related to a reconstruction and a re-assemblage focused on service or services. This reconstruction requires the redefinition of economic

7 The author owes much of this chapter to a presentation by Didier Chauveau about the normalization of the ISCC group, which took place on May 20th, 2009.

conditions, in particular the place of Fordism, in which the State had a major role. It teaches moving from the user to the citizen. It generates clashes of professional cultures; in other words, it is brutal. It is a way of thinking in reverse, as suggested by Benjamin Coriat.

Let us take the example of health. Com communication 689 (2008) is an example of willingness to change. It seeks greater efficiency and quality of care, associated with a reduction in costs and an increase in performance in health-related acts. This communication proposes to set up a new functioning model for organizations. It takes into account the financing of health and now a continuity in medical care, which is continually becoming more complex. Acts will increasingly come from communications between doctors, between the doctor and the patient and between the doctor and support services, which themselves will be more technical.

10.13.2. *The norm and machines*

The basic question is intimately related to language, and its answer is uncertain. The core of collective intelligence is outside, and it is in the objects. Therefore, it is a question of knowing what makes the norm to continue in this relationship between institutions and machines. Christian Lemoen proposed the idea that institutions play a role in keeping a record and reproducing norms. Thus, norms might be conceived as devices or crystallizations. It is an approach identical to that of Thévenot and Eymard Duvernay, who perceived institutions as "processed" piled layers.

According to Lemoen, projects mainly have represented forms of crystallization, and he pointed back to Leonardo da Vinci's "Designo". The ability to register projects in the materiality of social forms is achieved through norms. This implies several issues: first, there is the ability to focus on projects, but also being able to manage critical events, as well as the ability to anticipate breakdowns and produce traces. A norm is a universe; it is a holistic view of the world. Thus, projects that fail are those that are inscribed within a hereditary institution, which refuses any change in norms.

10.14. The impact of change on data

Let us recall that the issue of data is present from the very beginning on the instructions for tendering in the example of the CNRS:

> "Apart from scientific production, which is object of science itself, any industrial production, any commercial transaction and more generally, every human activity generates data which acquire the status of stored and exploitable information and knowledge, as far as digital automatisms are responsible for its management."

Project logic is introduced, and it requires cooperation. Thus, efficiency depends on mutualization. It represents the passage from the model logics to that of the process. The main change is that of the status of information. In this way, data about the organization involve moving from a logic of dissemination and transmission to a logic of access and publicized information. Data become shared resources.

Let us take the example of health again. The same applies to resources. The NATO COMEDS Group proposed incorporating telemedicine into its strategic axes, with a new approach, in which the STANAG would play an important role. Medicine is operating on a more and more cooperative basis, imposing real teamwork. Cooperation now goes beyond simple boundaries: doctors from a country hospital can accommodate fighters from another country, based on their mutual skills (as the current example from Afghanistan shows). This collaborative work is also enhanced by the development of advanced medicine and the delicate need for specialization in health professionals.

10.15. Changes in programs and processes

Norm is responsible for organizational changes, which have an impact on computer programs and processes.

It is interesting to observe that continua forum can self-add skills for creating reference computer programs and marketing intelligence, something that requires collaboration and mutual information, and finally, a testing activity and a certification.

The European Commission contemplated norm as a tool for change for putting into practice the computer platforms it deemed necessary[8]. It stated that Member States had to propose strategies for late 2009 and organize a conference of ministers on that topic in 2010. An assessment of telemedicine thus became possible by the end of 2011. It also proposed establishing norms by the end of 2010, especially for the interoperability monitoring systems. To do this, it proposed working on a white paper around interoperability and reminded that quality and safety were critical elements in normalization. In 2018, in different European countries, the administrative tools for the implementation of telemedicine are still not in place. At the same time, it started a research program: "ICT for health, aging and inclusion". This was proposed in the context of the information society policy. The proposed financing source, reaching the sum of 15 million Euros, was vested in two computer pilot projects "A" and "B". Pilot project A was conceived for being deployed on a large scale in Europe. Research was supposed to test the existence or the possibility of a personal health system in cardiovascular, chronic lung obstruction and diabetes areas. The Commission aimed at creating between six and eight healthcare providers.

10.16. Organizational evolution

In this context of full transformation, the norm will translate ongoing or future evolution in organizations and trades. This poses the question of the passage of actors from a participatory to a contributory mode. This last dimension brings together many questions, such as co-production, which is not always a founding principle, despite the fact that UGC (user-generated content) is often evoked. In organizations, the "functionalist Yalta" is disappearing; thus, DSIs and HRDs are becoming endangered species. The question of participatory innovation involves standardized data management and project management.

8 Telemedicine is an opportunity according to the Commission. It is the first time that the Commission has become truly interested in health. Indeed, the Commission is not competent on health, but competent on information technology and believes that telemedicine has strong export potential.

In this regard, we should not ignore the question of collaboration versus cooperation, dear to Marcel Mauss. On the contrary, it is clear that the terms of collaboration are being defined through norms that promote greater involvement from consumers or users in the production of goods and services. It is possible to speak of a prevailing trend, according to which collaboration is reflected in the market. In this case, the distributed expertise of the users is used for the production of the good or service.

Let us apply this approach to health again. The mHealth alliance resorts to cooperation when it announces it wants to develop a new care technology by working with partners to obtain the data necessary for solving problems and for improving communications during natural disasters.

First of all, this shows that there is no perfect organization.

Health online is an example of this knowledge industry. This document does not deal with the influence of standardization on the fluctuations between knowledge production, distribution and retention, nor on the value chain that it engenders. Should we integrate it into a normalization process or does this involve other approaches?

E-health represents the epitome of an industry that creates traces. Heated debates on the various forms of computerized medical records in different countries attest to controversies on digital intellectual property, which should be analyzed in depth. Even the name of the tool has been subject to debate. These digital traces are useful for the knowledge of healthcare professionals and hold important values for activities such as insurance. These digital traces can also be sources of information for other actors, for example, for future employers. The question now is what needs to be translated into the normalization process.

"Remote monitoring" is necessary for connecting the hospital with home. Indeed, the management of hospitals will have to reinvent itself. It is similar for the introduction of new care technologies in health professions. Professional activities that fall within this industry will surely emerge, and new workers on the knowledge industry will have to position themselves in the face of a world of management and care professionals. It will also be necessary to train them. The question here seems to be whether the resolution of these issues will undergo a normalization process.

10.17. The challenge of generating trust

The issue of trust derives from the emergence of marketplaces. Some are specialized in one area, while others are open to many products. But the challenge is also present in financing. Confidence in a brand is often very effective, but not always.

10.17.1. *Specialized marketplaces*

Artsy is a marketplace that specializes in art, which was created in 2009. An outstanding place for selling works of art, Artsy formed partnerships with art galleries and art fairs. In 2016, it established partnerships with big houses like Christie's or Phillips bids. Works of art are listed on the site. It is possible to buy or sell them. It is possible to learn about the artists and their works. The company boasts 20 million dollars of sales per month. Altogether, these are more than 1,800 companies, grouped in 90 countries, which list their objects on the platform, because these are gallery owners, art fair organizers and auction houses who register the art pieces, not the artists.

10.17.2. *Rating, the representation of trust*

In order to create trust, the actors of society 5.0 transfer confidence to the relationship. In fact, they encourage the user to appreciate the relationship they have with each other, rather than the one they have with the platform. Thanks to this, a miraculous solution became generalized: rating. The key tool, rating, should be mutual as far as possible, in order to grant the seriousness of confidence. The rating system is not new, every student receives marks, and old TV shows, such "Fan school", or, more recently, fashion or variety shows use ratings. In other words, ratings are not new in our daily lives. However, rating shifted from social rating, which served as a ranking:

– to the economic impact on platform users;

– to an analytical tool for the owners of platforms;

– to a possible governmental control for monitoring certain organizations.

It is precisely because actors need trust that ratings become decisive and a source of various biases.

10.17.3. *Commitment as an ingredient of trust*

Customer commitment toward a service platform and the commitment of an employee toward an organization are two forms of trust that should be implemented in society 5.0. Commitment is a key concept of this type of society.

DEFINITION. Commitment means to trust actors with a mission, to keep their motivation and even to develop it.

Training is made up of the following four steps:

– the communication of examples and inspiration;

– the co-development of solutions (this stage is entrusted to people coming from different functions or origins and it entails building new landmarks rather than developing readymade solutions);

– the contribution of tools for implementing specific benchmarks;

– the facilitation of the implementation, and as well as the facilitation of things.

To embark people on an adventure, it is necessary not to immediately set the goal too high. Finally, the famous sentence by Saint-Exupery, seeking to engage commitment, is appropriate: "When you want to build a ship, don't start by gathering wood, cutting planks and distributing tasks, but awaken within men the desire of the big and beautiful sea."[9] The idea is to propose an easy first step, so as to avoid resistances, be inspirational and promote deployment.

10.17.4. *The necessary confidence for inviting financing*

The traditional functioning of banks and credit agencies is wrongly set in the digital world.

The digital world is largely immaterial. Banks and credit agencies want to take physical grant in order to secure their loans. It is therefore easier to finance facilities or machines, rather than software. Trust should be established in this type of purchase between the backer and the buyer.

9 Excerpt from *La Citadelle* by Saint-Exupéry.

The digital appears as a charge in international accounting regulations. Investment in the digital does not appear as such in the balance sheet. This affects the analysis of the strength of the firm, even if this has become more efficient. Trust rather will depend on the accounting valuation of this intangible patrimony.

Proving the profitability of a project and its return on investment is almost impossible, because we lack achievement references, in virtue of its innovative character. It is also difficult to assess the evolution of the functioning of business and possible revenue according to these new investments.

"Servitization" associated with digitization is inevitable. It follows that the level of investment is difficult to assess, and will probably increase, which intensifies the problems of funding. To this, we should add the financing of physical objects (including machines) as well as the premises for housing the staff.

The effects of branches are also important. A large digital firm will impose its solutions to its subcontractors, who do not generally have the same financial means, resulting in critical funding issues. One idea would be to define sector strategies in order to reassure financial institutions.

In every country, it is necessary to build institutions capable of answering these needs. Financial institutions should become used to the culture of the digital customer and reorganize themselves to meet the needs of organizations. Luckily, digital is not expensive and can be self-financed.

11

On-Demand Society

On-demand society finds its origins both in "servitization" and in a response to boredom. Society 5.0 is essentially an on-demand society, and this demand has not found a spontaneous reply from productive organizations. They were obliged to adapt themselves.

11.1. Does boredom have any influence on need, desire, expectation and demand?

At this stage of the analysis, it is necessary to replace boredom in the chain need–desire–expectation–demand. We have to analyze its diverted uses, which may be a response to boredom in one case, or the expression of a real need, in another.

11.1.1. *Collective neurosis and diverted uses*

In his work *Malaise dans la civilisation* Freud stated the hypothesis that collective and individual developments are coupled. The individual is structured by collective neurosis according to three identification levels. At the family level, the individual is inscribed in the order of generations. At the group level, the individual acquires a social identity associated with the groups to which it belongs (school, profession, social class). The individual acquires a collective identity, which includes two main dimensions: a real dimension related to history and an imaginary dimension related to culture. When collective neurosis works harmoniously, these three levels are coherent. Liaudet introduced two rules[1]. If individual neurosis is the only reflection of collective memory, then happiness depends on the latter. If, as a result of historical and cultural changes, the collective neurosis becomes disorganized, then it might promote the appearance of pathological symptoms in individuals. These symptoms may then be projected on the collective neurosis and transform it.

11.1.2. *The theory of diverted uses and the role of boredom*

The research we have carried out shows three levels of diverted uses. Some consultants even use the term "unforeseen use". From the marketing perspective, the sole interest of diverted uses is to increase the volume of consumption of a product or service.

The diverted use of the first kind appears when we exacerbate one of the functions of the product. For example, the mobile phone is a phone with an extra function related to mobility. If you are driving and you find yourself in the middle of a traffic jam, already late for your meeting, the idea of stopping to phone from a public phone is not adequate, because you would still lose more time. On the other hand, with your mobile phone, you can call from your car. Therefore, you will increase the use of your mobile phone. This situation might be perceived as a form of melancholy or grief related to the event of a loss of a commercial prospect or even the relationship with friends. Only a diverted use of the first kind is a response to consumer satisfaction in the terms previously described. Our consultant Nathalie Joulin focuses mainly on this logic. The new "innovation race" comes from a "more discreet science", with "heightened functionalities"[2].

Boredom is the source of a diverted use of the second kind. Let us use the example of the mobile phone again. Let us imagine that you are waiting for the bus, when boredom sinks in. Regularly calling your mother-in-law is also a source of boredom. But you decide all the same to call her while you wait for the bus. She is happy, you are too, because you spent your time usefully, your wife is also satisfied. Finally, the arrival of the bus is a great opportunity that will allow you to finish the communication.

1 Jean-Claude Liaudet, *op. cit.*, p. 23.

2 Nathalie Joulin, *op. cit.*, p. 13.

The diverted use of the third type is related to situations where anguish and anxiety are involved. Let us imagine ourselves in the skin of the young mother who has just had a baby; the idea of going to the theater with her husband is unbearable because of the anxiety to leave her baby. Let us equip our young mother with a mobile phone, anxiety finds its remedy in anguish. She will accept to leave her child with someone to babysit. Mobile phones make it possible to confront anguish, despite the ban on using phones in the middle of a presentation. The young mother will call the babysitter three times, once before the beginning of the show, to inquire about whether everything is going well, at intermission for the same reason, and also at the end, to inform about her return. Three additional calls have been made.

11.1.3. *Examples of diverted uses*

Some examples of diverted uses are famous in managerial literature. Due to their performance, in general, they have been dethroned by other devices. Table 11.1 shows some of the best-known examples.

Products	Diverted use
Graham Bell phone	Device for communicating from person to person Diverted as a phone theater, it was later used for listening to performances at home Finally dethroned by radio and television
Thomas Edison phonograph	Device for transmitting professional messages Diverted for listening to music Dethroned by the vinyl record, and then the DVD
Mobile phone SMS	Used by the phone operators for sending information to subscribers Diverted for sending messages between subscribers Dethroned by smartphone applications

Table 11.1. *Examples of diverted uses*

Drugs are often diverted by the end user with the prescriber's approval. Let us mention Viagra, which had been designed as a drug for heart disease and whose secondary effect was put forward, to facilitate erections. The mediator which was originally an admixture for diet in diabetic patients had become an appetite suppressant.

A collective work, conducted by Allard, Creton and Odin[3], focused on mobile phones and their diverted uses. Some developed theses are interesting for explaining diverted uses, such as the one by Maurizio Ferraris, according to whom "the mobile phone is a typewriter"[4], or the one by Serge Tisseron, for whom the "mobile phone has, of course, revolutionized photographic practice"[5]. Serge Tisseron analyzed the triple revolution brought by the mobile phone: liberation of the practice, possibility of immediately seeing the result and the reversal of the relation between photography and the act of seeing[6]. A less popular thesis, that of Roger Odin, Benoît Labourdette and William Uricchio, considers this device as a cinematographic tool[7]. William Uricchio recalls, "in general, the history of cinema is especially rooted in the documentary reflex"[8]. Furthermore, he explains, "from its very beginning, the documentary connected viewers with world affairs."[9] Through these examples, it is easy to imagine the magnitude diverted uses have acquired.

11.2. "Servitization", the products and services of revolution 5.0

After servicing, "servuction" and the fabless, society 5.0 is characterized by the promotion of "servitization". The service society in which we find ourselves is undoubtedly linked to the slowdown in product innovation. It is for this reason that the actors from industry have engaged in the "servitization" of products and production. This development has certainly prompted a major change in our societies: instead of selling a car to a private individual or a train to a firm, which will subsequently transport users, it is the transportation services from one point to another that are commercialized. Thus, we can also speak of a "service industry". This evolution has resulted in changes in the dominant players and in the importance of the role of platforms. This requires these actors to change their practice when new competing actors are emerging.

3 Laurence Allard, Laurent Creton, Roger Odin (ed.), *op. cit.*

4 Laurence Allard, Laurent Creton, Roger Odin (ed.), *op. cit.*, p. 23.

5 Laurence Allard, Laurent Creton, Roger Odin (ed.), *op. cit.*, p. 117.

6 Laurence Allard, Laurent Creton, Roger Odin (ed.), *op. cit.*, p. 118.

7 Laurence Allard, Laurent Creton, Roger Odin (ed.), *op. cit.*, pp. 37–79.

8 Laurence Allard, Laurent Creton, Roger Odin (ed.), *op. cit.*, p. 65.

9 Laurence Allard, Laurent Creton, Roger Odin (ed.), *op. cit.*, p. 66.

11.3. The notion of "servitization"

"Servitization" can be explained by a three-item diagram, as illustrated in Table 11.2.

Type	Description
Production	Search for performance in production Strengthening the firm, "servitization" is perceived as a fundamental production factor
Organization	Having the means for developing institutional architectures adapted to future industries
Products and services	Combining products and services systems by using information and communication technologies with a worldwide aim

Table 11.2. *The logic of "servitization"*

DEFINITION. "Servuction" is a term that specifically refers to the production of services.

The term is a neologism constructed on the basis of the words "service" and "production". It was developed by Pierre Eiglier and Éric Langeard and appeared in the 1980s[10]. This approach was later used by Badot and Cova[11]. "Servuction" differs from the classic production activity in that its essential component is marketing, which might account for the fact that marketing appropriated it. It essentially concerns the relationship with the customer, whereas the production of a classic good is essentially an "industrial" problem. In fact, nowadays, Client Drive production units are evolutive forms of "servuction". This approach has been included in the marketing of services. The term "servuction" is currently fading in its uses, to the benefit of "services design", that is, services that include a material component.

11.4. The nature of "servitization"

The nature of "servitization" lies in innovation and the design of services, a concept that still differs from servicizing and "servuction".

---

10 This refers to the work by Pierre Eiglier and Éric Langeard, *Servuction. Le marketing des services*, 7th edition, Ediscience, Paris, 2000.

11 This is the work by Olivier Badot and Bernard Cova, *Le Néo-marketing*, Editions ESF, Paris, 1992.

11.4.1. *Servicizing*

DEFINITION. In managerial literature, servicizing has simply been defined as "selling a service rather than a product" (Makower, 2001) or as "the sale of a function of a provided service or product, rather than the physical product itself" (Fishbein *et al.*, 2000).

Leasing a product whose use we are renting is the example suggested and studied by Fishbein. For these authors, the main advantage of servicizing is the reduction of transaction costs (transaction cost economics or TCE). All the studies which have embraced this approach focused on the two possible scenarios in which servicizing rather originated in firms or organizations which were seeking a reduction in transaction costs, or, on the contrary, from consumers lost in multiple offers.

11.4.2. *The different forms of servicizing*

Regarding service in particular, an acquisition model based on the rental of services, instead of their purchase, appeared. The business model then started being based on pay per use. The access model works on the Internet on any terminal. The technical component is defined as on-demand, "scalable" and shareable. Some years ago, this form used to be sold under the name of "service desktop". It has recently been renamed "cloud computing". Different factors have favored the emergence of this new model[12].

Factors	Comments
Regulatory	The emergence of the Internet of services, a new Web standard
Technological	Virtualization (bug tolerance), importance of networks (availability, throughput, latency), feeble latency of communication technologies (fiber and 5G mobile)
Economic	Economic crisis, rivalry, competition, sustainable development
Uses	"Anytime anywhere and any device" (ATAWAC) access, pay per use

Table 11.3. *Functional characteristics of cloud computing*

12 This section is inspired in the book by Bruno Salgues, *L'E-Santé et la télémédecine*, Hermès-Lavoisier, Paris, 2013.

A new difficulty came from the risk of "pipe fill-up". In this case, a popular saying ought to be remembered: "Expose simplicity, hide complexity". The following three types of solutions thus became possible:

– Software as a service (SaaS), based on applications, services and app stores;

– Platform as a service (PaaS), which provided the tools;

– Infrastructure as a service (IaaS), which offered computing and storage capabilities.

11.4.3. *"Servuction"*

By creating the term "servuction", Eiglier and Langeard (1987) adopted a systemic conception of service production. Thus, service should be understood as the goal and the result of a provider's "servuction" system. Therefore, "servuction" can be defined as the systematic and coherent organization of every physical and human customer–firm interface necessary to the delivery of a service, the commercial features and quality levels of which have been predetermined. When we are interested in a service, but not in a product, it becomes necessary to make a distinction between the main service (core service), the one which satisfies the basic need of the customer, and the auxiliary or supplementary service (peripheral service). All these services facilitate access to the basic service or add value to the defined service (Eiglier and Langeard, 1987 or Grönroos, 1984, 1993, 2000).

11.4.4. *Competitive advantage*

Michael Porter has defined a theory of competitive advantage taking the notion of value chain as a point of departure. From this, he deduced the fact that there are only three ways to build competitive advantage: cost reduction, differentiation or focus on a niche (which we previously discussed) and the creation of new industrial competitiveness.

The differentiation strategy seeks to obtain competitive advantage through differentiation. Thus, it is possible to offer a product or a service with high apparent added value to a broad target, with small volumes, but strong margins. This is the very essence of the "servitization" strategy, which is no more than positioning oneself closer to consumers in response to their needs.

Cost leadership strategy is used within the “servitization”, resorting to a counter-use. In order to achieve cost reduction, there are also use restrictions. A consumer who uses his car very little can rent it when he wants to. If his needs are relatively low and he uses his car only during peak use times, the solution of the rental will look cheaper for him. In this case, the cost reduction strategy is quite different from a low-cost strategy. In this way, the competitive advantage is obtained by offering a product at a low cost to a strategic target, with small margins, but large volumes.

The approach through concentration is often implemented through the multiplication of distribution channels. The focus strategy aims to gain advantage by offering a product or a service to a small target, whether this is locally delimited or small in terms of product functionality. Accordingly, we can perceive the adaptation of product discourse to different audiences. In order to achieve this, it is necessary to adopt a chain of values adapted to the expectations of the target.

Everything else is just getting stuck on the middle way and probably leads to failure.

11.5. The paths toward “servitization”[13]

“Servitization” is the result of some recent, established findings in the field of the use of technologies. The first observation is that many sectors have reached their limits in cost reduction. The second observation is that it has become difficult to sell hardware through differentiation, which imposes the need to offer added value by means of services. Information and communication technologies (ICT) have a high capacity, and they make it possible to easily add these services to products.

13 This section owes a lot to my discussions with Yoshio Tanaka, Professor Technology Dynamics at the Department of Management of Technologies (MOT) from the University of Tokyo, on the occasion of the Global Forum “Shaping the Future 2013, Driving the digital future, strategies to grow stronger communities and businesses”, October 28th and 29th, 2013, *Palazzio dei Congressi della stazione Marittima*, Trieste, Italy.

11.5.1. *The formation of value*

The idea is to become interested not in the decomposition of value as Michael Porter did, but in the formation of this value, even if it means reprising a Marxist approach. So far, value has been generated through costs, usability, satisfaction and, more recently, by the "green" side. Therefore, it is important to think about the process of introducing services, not in terms of "servuction", as we did previously, but in terms of "servitization".

How is value created?			
Costs	**Usability**	**Satisfaction**	**"Ecology-friendly"**
Reduced costs by sharing infrastructure	Business agility	Stress reduction through greater consumer participation (co-creation)	Reduction in the consumption of certain resources (sustainability)
Knowledge sharing	Quick start, quick response to demand	Participation in development communities (co-development)	Knowledge sharing
Reduction in the costs of production capacities	Security	Co-construction of the product with the customer	Green washing or traditional washing[14]
Reduction of operational costs	Data governance (Open Data, etc.)	Deletion of physical relations (ATMs)	Change in manufacturing inputs and accepting an increase in costs
Frugal innovation	Social innovation	Techno-marketing innovation	Environmental innovation

Table 11.4. *The formation of value*

This value formation table presents four types of innovation of society 5.0, namely, frugal innovation, social innovation, techno-marketing innovation and green innovation.

14 This means giving the impression that the product and associated services are "green" or traditional. In 1992, Greenpeace published the *Greenpeace Book of Greenwash* in order to denounce these "marketing" operations.

11.5.2. *"XaaS" logic*

In the firm or within organizations, "servitization" is carried out in XaaS mode ("X as a Service") .

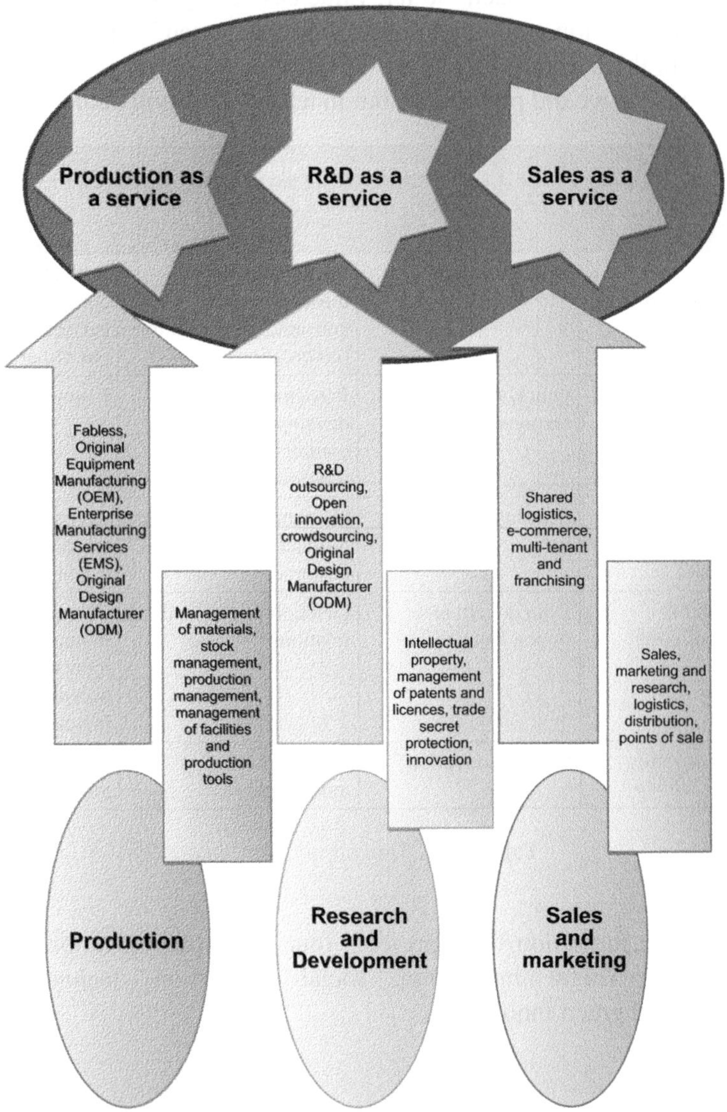

Figure 11.1. *The formation of XaaS in "servitization". For a color version of the figure, see www.iste.co.uk/salgues/society.zip*

Although it is true that traditional sales have existed for a long time, which is reflected by our downtown stores and mass retail, other manifestations are more recent.

11.5.3. *The "rental" rather than the "purchase" logic*

A new "servitization" economy has developed thanks to the professionalization of collaborative economy. Carpooling, urban bike rental and efficient technical solutions have disseminated over other products.

Rental introduced novelty and thus became a tool for fighting against boredom. It is possible to rent a personal daily wardrobe. Without the need for libraries or furniture for storing DVDs, the e-reader now enables us to read and streaming or digital channels make it possible to watch movies. In Japan, it is possible to rent dogs for walk in public gardens.

Rental has shifted from an occasional need to a daily need. Television rental companies offered occasional entertainment during the hospital stay, in the same way that DIY equipment could be hired at Loxam or Kiloutou. In the same vein, it was possible to hire a ceremony vestment for a special occasion. Car rental became a habit thanks to long-term rental offerings, and this methodology is spreading to furniture and large appliances.

Two logics complement each other in this approach to rental, apart from being an answer to boredom. First, there is the logic of availability; equipment that breaks down or rapidly becomes obsolete, is quickly replaced, without waiting for a possible repair. Rental may be perceived as virtuous, because repaired devices can easily find a second life, especially on a second-hand market, which might also serve a social purpose, by favoring the access of less well-off populations to these devices. This has been checked for vehicles and electronic devices. The second logic is that of immediate pleasure. The device is immediately made available, and the product in which one invests does not devalue as quickly as a car. The same goes for clothing, where it is possible to "wear brands" and regularly change the look[15].

15 As an example, let us quote the following website: https://www.lhabibliotheque.com/.

According to Lokeo[16] platform, its customers are young people between 30 and 40 years, belonging to superior social categories and who are early adopters. This platform claims 50,000 customers in France and is constantly expanding its range of products available for rental and additional services, such as subscriptions.

According to these new approaches, rental is more a one-off and exceptional need, permanent use of which might correspond to "servitization", a typical feature of society 5.0.

11.6. Enterprise manufacturing services

Enterprise manufacturing services (EMS) developed with the growth of fabless or firms without a production logic, whose French promoter was Serge Tchuruk.

11.6.1. *The fabless*

The term fabless was launched in June 2001 by Serge Tchuruk. In a conference organized by the Wall Street Journal, the number one from Alcatel[17] launched the term fabless in order to describe the future of his company. This contraction between the English terms "fabrication" and the suffix "-less" was astonishing at the moment and described as a "bomb". This concept of an enterprise without a factory was extremely popular at a given point and is currently fashionable again.

Fabless logic has been criticized and presented as a somehow disguised form for large groups to avoid laying off a lot of employees. The manufacturing unit of Alcatel mobile phones, located in Laval (Mayenne) in France, was sold in 2001 to Flextronics. At that moment, this unit had

16 Available at: https://www.lokeo.fr/.

17 Alcatel, now Alcatel-Lucent as a result of its merger with the American in 2006, later bought by Nokia, has nevertheless preserved factories, for example, at Eu (Normandy) and Calais (North). Before the plan, the firm used to have 80,000 employees worldwide, against more than double in 1995.

830 employees. The decision to close the site was presented to the Council in October 2005, and the layoffs were completed throughout 2006. The former factory of Matra Nortel Matra, and Nortel in Châteaudun, was eventually taken over by Flextronics in August 2005. Jabil resumed the production of Alcatel PABX in Europe and acquired the industrial site of Brest in 2002, which had 1,200 people. On the contrary, the plant itself had only 185 employees.

11.6.2. *Original design manufacturers*

Original design manufacturers (ODMs) carry out design, validation, distribution activities and/or offer after-sales service. More recently, EMS have started offering additional services. Some EMS made the choice to reconvert into ODM, thus becoming full-fledged or fully "owned brands"; as an example, we can mention two Taiwanese firms: ACER (computer science) and HTC (mobile phones).

11.6.3. *The example of the EMS of electronics*

Regarding EMS, they mainly focused in electronics because of the rise of Asia in these areas.

In this case, EMSs act as electronic manufacturing services. We can differentiate two kinds:

– Printed circuit board manufacturers, which only manufacture printed circuits, and

– PCBAs, which produce the assembly of printed circuits and sub-assemblies. The customers who command the jobs are original equipment manufacturers (OEMs).

Currently, these electronics firms are an important part of production chains, which use electronics.

The best-known electronic manufacturing services are:

– Actia Group

– All Circuits (from MSL and the EMS activity, from Sagem Communications)

– Asteelflash Group (France)

– Benchmark Electronics

– Celestica

– Enics (Switzerland)

– Eolane (France)

– Flextronics (which bought Solectron)

– Jabil

– Lacroix Electronics (French number three)

– New Kinpo Group

– Plexus

– Sagem Industries

– Sanmina-SCI

– Selha

– Shenzhen Kaifa Technology

– Tronico

– Universal Scientific Industrial Co., Ltd.

– VIDEOTON Holding (Hungary)

– Zollner Elektronik Group (German leader)

Box 11.1. *Electronic manufacturing services*

11.7. The key points of "servitization": visualization and virtualization

In this section, we will deal with the key points of "servitization", as well as with the current developments in this area.

The two main points of "servitization" are visualization and virtualization, two techniques related to information and communication technologies.

The diagram in Figure 11.2 offers the necessary tools for "servitization". Virtualization, which has been involved in the servicizing process, is not sufficient to set up "servitization".

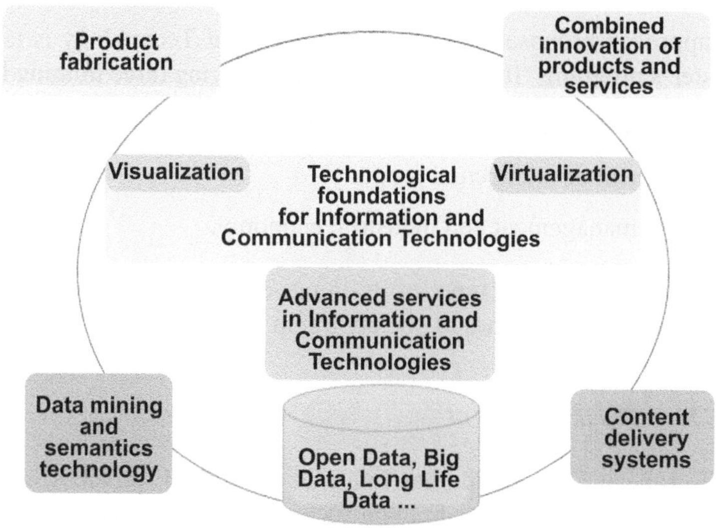

Figure 11.2. *Use of ICTs for "servitization". For a color version of the figure, see www.iste.co.uk/salgues/society.zip*

For this, it is also necessary to have visualization and data management processes, where the latter decomposing technology is related to data mining and content delivery technologies.

Here, we will not discuss the emerging logic differences that we perceive between Open Data, Big Data and Long-life Data.

11.8. Recent developments

In order to develop "servitization", the Japanese have created a study group gathering industry representatives, consultants, academics and members of the administrations around Tokyo University of Technology.

The idea is to reflect upon the means to facilitate innovation. This group concluded on the importance of the development of information and communication technologies (ICT), as well as their associated infrastructure. The European Commission also became interested in this topic. This chapter includes some examples of these developments.

11.8.1. *Tokyo University of Technology*[18]

The approach proposed by Tokyo University of Technology is taught in two Master's programs. It is divided into the following three training blocks:

– management of technology and its dynamics;

– management of intellectual property;

– business management and industrial economy.

The example of this type of training enables us to appreciate the importance of the approaches of society 5.0 in Japan.

11.8.2. *The SPREE project*[19]

The European Commission has financed the SPREE project (Servicizing Policy for Resource Efficient Economy) as part of the Seventh Framework Program (FP7), a project which analyzes how to move from a society selling goods to a partnership for the supply of services. One of the conclusions of this group confirms the vital role of infrastructure as well as normalization policies.

11.8.3. *The example of the firm Komatsu*

The example of the firm Komatsu[20] is emblematic, because it has enabled this company to become a leader in its sector and, consequently, to dethrone Caterpillar.

18 Tokyo University of Science, Graduate School of Innovation Studies, Department of Management of Technology (Word), PORTA Kagurazaka 2-6 Kagurazaka, Shinjuku-ku, Tokyo 162-0825. Available at the following website: http://most.tus.ac.jp; email: mot@admin.tus.ac.jp.

19 SPREE project (*Servicizing Policy for Resource Efficient Economy*) available at: http://www.spreeproject.com.

Komatsu has equipped its trucks and excavators to monitor remote operations and enable the management of the fleet. The offer named "Komtrax" is a mobile application, compatible with iPhone, iPad and Android smartphones. It makes it possible to manage a fleet of mobile equipment of the firm (trucks, excavators, construction equipment) remotely, in a proactive way and at any time.

Among the data included, let us mention the hours of use (SMR), fuel consumption, alert codes, especially breakdowns, travel and route tracking, as well as GPS location.

20 Komatsu is a registered trademark. KOMTRAX is a registered trademark of Komatsu America Corp., an authorized dealer of Komatsu Ltd.

12

The Economy of Society 5.0

The economy of society 5.0 is characterized by not only a search for solutions to the problems of economy 4.0 but also the birth of new forms of economy. It has been known for 40 years thanks to many econometric studies that there is no relationship between the expenditures in research and development (R&D) and the economic performance of firms: profitability, return on investment, growth. The relationship between spending on R&D and performance seems easier to perceive, where R&D produces direct impact (the case of a positive relationship between expenditures on R&D and gross profits), but if we add the costs of marketing, sales and administration, the relationship disappears. On the contrary, a large part of the costs of the product can come from R&D, as it happens in the case of the drug industry. The link between R&D and performance is no longer visible, and this situation requires making R&D more efficient and/or more effective.

In order to make the organization more efficient, there are only a few methods:

– placing oneself in a new economy and taking profit from novelty;

– developing a good business model;

– implementing capabilities and skills related both to cross-functional data and related to the outside world.

Although it is true that the first form previously existed, the notion of business model first appeared as a response to the end of major innovations. Osterwalder (2004)[1] dated the first occurrence of the term in the 1960s, in an

1 This makes reference to his thesis at the University of Lausanne in 2004.

article by Jones[2], which was difficult to find for long. The oldest and most convincing definition of the concept was proposed by Viscio and Pasternack in 1996[3]. The third solution, that is, the most recent one, is characteristic of society 5.0.

12.1. The new economies

Different forms of economy are contemplated. Table 12.1 presents a few possible scenarios, which can be added up.

Name	Features
Economy 2.0	Economy based on information and communication means
Blue economy	Strong growth and high profits that a firm can generate can only exist by creating a new application in an unrivalled strategic space, called "Blue Ocean"
Circular economy	This economic concept is part of sustainable development. The goal is to produce goods and services while limiting the consumption and waste of raw materials, water and energy sources
Collaborative economy	Taskforce sharing type of economy
Design economy	Highlighting the importance of training, functions and conception
Economy of distraction	Economy related to theme parks and tourist attractions
Exchange economy	A type of economy in which one seeks to increase exchanges
Gig economy	Economy dominated by new modes of work, including freelance work, micro-jobs and project contracts, is usual
Free economy	Economy in which the consumer is separated from the payer who bears production costs
Gray economy	This is an underground economy, in the sense that it is not controlled by governments
Economy of intangibles	A form of economy that is digital and therefore removes the forms of previously existing material exchange
Economy of indignation	A form of economy in which people feel indignant about certain topics in the hope of collecting consent, but which leads to outbidding favorable or unfavorable to the sales of products and services

2 This corresponds to the article by G. M. Jones, "Educators, Electrons, and Business Models: A Problem in Synthesis", *Accounting Review*, vol. 35, no. 4, 1960, pp. 619–626.

3 This is the article by A.J. Viscio, A. Pasternack, 1996, "Toward a new business model", 1996. Available at: http://www.strategy-business.com/press/16635507/14974.

Name	Features
Purple economy	It is a type of economy that adapts to human diversity in a context of globalization. It relies on the cultural dimension so as to attribute value to goods and services
Sharing economy	A form of economy in which the limits are respected and in which players share production or resources
Economy of proximity	This type of economy is based on the relevance of short distribution channels
Quaternary economy	A type of economy that combines the secondary sector and the tertiary sector whose products are neither goods nor services, but "new services incorporating goods, with a temporary provision of goods, people or combinations of goods and people". The quaternary sector also comprises information and communication (ICT) technologies, as well as the training associated with them, and governmental actions to support their development
Retro economy	Economy employing retro aspects and the promotion of products
The economy of services	A type of economy in which property is strongly reduced in the face of rental and "servitization" is widely implemented
Silver economy	Economy based on the growth in the number of elderly people and their needs
Solidarity economy, Social economy	Social economy or social and solidarity economy (SSE) is the branch of economy that reunites private organizations (firms, cooperatives, associations, mutuals or foundations) seeking to reconcile economic activity and social equity
Slow economy	A type of economy that struggles against excessively fast effects
Startup economy	Economy based on innovative firms, usually looking for significant fundraising, with a very strong potential for economic growth and and financial speculation regarding the future value of the firm
Economy of uses, Economy of experiences, Economy of functionalities	Previously, economy used to be based on the sale of a good requiring the use of a service. Now, a service or an integrated solution is sold in order to meet the same functionalities as the good previously sold alone, while consuming fewer resources. In addition, in this regard, the functionalities of the product can be expanded.
Green economy	Economy which produces improved human well-being and social equity at the time that it significantly reduces environmental risks and the shortage of resources

Table 12.1. *Different types of economy*

At this stage, we should note the multiplicity of savings described as a part of society 5.0 and the dismay of traditional economists when they have to describe the economic features of this new type of society.

12.2. The problems in the age of connectivity

The machine has connected actors among themselves and has not produced the same effects everywhere. In fact, the expected link between technique and social progress is no longer real. At the micro-economic level, productivity gains are not significant and growth remains weak. Education and health are the two sectors that have been least impacted by the economy, which was born with information and communication technologies.

Connectivity has not significantly affected democracies where they exist. Still, we should note the existing willingness to use digital technology in order to implement the necessary means for moving toward direct democracy. The world of connectivity affects certain States in which democracy is not so strong without this process being extended to all non-democratic nations. Development in these countries as well as in developing countries is still associated with the implementation of infrastructure and catching-up processes.

In order for productivity gains to emerge, it is necessary to redesign the production process of goods and services. For some, this is called "uberization", the name of a very famous firm. One of the observed factors of connectivity is the reduction of the possible size for an economic actor to emerge on the market; in other words, the entry ticket to the market is lower, which has an immediate consequence: the reduction of financial needs.

The old forms of economy will have to be abandoned in favor of modern forms.

12.3. Evolution of economy

Man in society has always had the desire to improve his lifestyle. For this, he evolved in the way he obtained goods and services he wished to use. These modes of acquisition have shaped the economies that have evolved over millennia. Curiously, in our current society, we can find examples of each type of economy, coexisting with the largely dominant market economy. Hence, society 5.0 corresponds to an accumulation of the benefits of each of the periods. The hunting and gathering period exactly corresponds to society 1.0. The agrarian period first corresponds to a bartering economy, which was replaced by an economy of souk with the birth of currency.

Types of society	Types of economy
1.0. Hunting and gathering	Hunting and gathering economy
2.0. Agrarian	Bartering economy
	Souk or bazaar economy
3.0. Industrial	Production economy
	Mass distribution economy
	Market economy
	Environmental economy
4.0. Information	Digital economy
	Intangible economy
Society 5.0	The platform unifies all previous types of economy (bartering, souk, production, mass distribution, market, environment, digital, intangible). Various forms of economy (see Table 12.1)

Table 12.2. *Different types of economy*

12.3.1. *Hunting and gathering economy*

The consumer/predator learns to orient himself through a jungle of products available in his/her environment. The selection criteria concern what is edible, nice, big or easily accessible (within hand's reach).

Depending on whether there is abundance or shortage, the consumer/predator will have easier or more difficult access to the products he/she desires. Nomadism is the rule that makes it possible to find new or missing products.

Nowadays, the concept of gathering is applicable to hypermarkets, in the goods as displayed within hand's reach.

12.3.2. *Bartering economy*

The consumer/producer becomes specialized and learns to exchange what he/she has in abundance so as to obtain the products he lacks.

This behavior has deeper meaning than sharing food and protecting the weakest of the clan. These exchanges are sealed by rites, and some archaic societies are still experiencing ceremonial bartering, whose meaning goes far beyond the mere contract. These exchanges are privileged moments of communication, during which man acquires objects/symbols.

Currently, all courses in sales techniques that take interpersonal relationships and neurolinguistic programming (NLP) into consideration have analyzed human gestures.

Bartering economy, still in effect in the countryside (exchange of a duck against fruit, loan of the lawnmower in exchange for the hammer-perforator), has been regenerated and found new applications thanks to information and communication technologies.

On the Internet, we can find bilateral or multilateral exchange platforms: A proposes his product to B, who proposes his to C, who proposes his to D, who has an available product or service that A needs.

12.3.3. *Souk economy or the basis of market economy*

The consumer learns to use the currency and to identify places devoted to trade.

Some well-known concepts thus emerge: display, auction, abundance and the choice of products that can come from all parts of the world. At the same time, the first windowless stalls are created, in which only the sign indicates the profession; inside the shop, the counter is in a privileged position, products are stored at the back of the store and the salesperson releases the goods he considers appropriate to display.

The salesperson/craftsman and the salesperson/importer rule and become an obliged passage between the customer and the product.

12.3.4. *Production economy*

At the stage of production economy, production means become concentrated and find a special location. Supply is lower than demand.

Factories become production plants and crafts give way to mass production. Distance from the place of production as compared to the place of exchanged gives birth to salespeople/representatives who contact the shops or who reach consumers door-to-door. Producers no longer control the sale of their goods, but they start advertising and consider that the product will somehow be present in the minds of traders, having gone through the intervention of representatives. At the same time, they make themselves available to customers. Emphasis is placed on improving production techniques and the scale becomes supra-local.

In *La Vie quotidienne dans le monde moderne,* Henri Lefebvre described the mentality of that time to us:

> "The theorists of Consumer Society [...] state that once, at the beginning of the capitalist economy and of industrial production, in the prehistory of modern society, needs did not steer this production. Entrepreneurs were unaware of the market and ignored consumers. They produced randomly, launching their goods on the market waiting for the buyer, expecting the consumer."

12.3.5. *Mass distribution economy*

When production finishes its structuring process, the possibilities of selling goods become huge. New distribution channels are born in the face of strong demand, and large masses of consumers are potentially affected, because the scale is supra-local.

Supply tends to match demand. Department stores, supermarkets and then hypermarkets are able to work with traditional distribution channels. With a proper organization of the parcel delivery, producers can even sell at distance, by mail (mail order), a type of sale that has now been enhanced by the Internet and the Web. Distribution is organized with the appearance of central contracting agencies and listing policies. The salesperson disappears behind the product and vanishes from self-service stores. Advertising becomes publicity.

Mass production is accompanied by mass distribution, and we are entering an era of consumption that seems exponential.

12.3.6. *Market economy*

This paradise of consumption, which has led to the expression "big is beautiful", is not endless. Purchasing power is not exponential, and competition is fierce. Supply is higher than demand.

High-profile bankruptcies oblige business leaders to reflect. They start by studying consumers, launching adapted products, that is, the marketing approach, and seeking to determine which products best suit consumer expectations.

Since then, studies have strongly expanded their fields of investigation. The concentration of production means and distribution continues despite the shift from uniformity to product customization (sharper consumer segmentation).

12.3.7. *Environmental economy*

Beyond the strict notion of market, businesses depend on other firms for their supplies, as well as on other markets. The internationalization and interdependence of all actors lead to an environmental economy.

Nobody is isolated in the market. Manufacturing a product or not, using a specific material, can lead a country's economy to trouble or can significantly pollute the planet, and decisions may depend on customs barriers or on the authorization of a simple municipality. The environment is unstable, markets are saturated, and information and communication technologies develop and pave the way to greater investigation opportunities.

Market analyses focus on the environment. Strategic marketing acquires a prominent place (the development strategy of a firm develops taking current factors and potential changes of the environment as a whole into consideration).

The success of a product comes from its overall environment, which can also be called ecosystem.

12.3.8. *Intangible economy*

Intermediation in the value chain is totally questioned by the development of information and communication technologies (ICT).

The market is no longer made up of producers on the one side, and homogeneous consumer groups on the other side (consumers who supposedly have the same desires, expectations and requests in order to satisfy their needs).

The market is made up of cultural groups, called "tribes", who can spontaneously enter into contact with the producer of a good or a service. Each member of the tribe can also contribute to the development and production of a product (collaborative production) or to its distribution by offering technical, organizational and human means to the other members of the tribe.

Producers may distribute, exchange and market their products directly or they may use new intermediaries.

The analysis of the market must take into account the specific construction of the value chain, by differentiating the customer from the consumer and the producer from the distributor. Market analysis can be made independently from the analysis of the economic model of these market players. A fixed and thorough description of the market becomes impossible at this stage.

The complexity of this economy makes the analytical approaches of traditional marketing difficult and requires a systemic approach, which can enrich the marketing techniques of previous stages, once the significant components of the problem have been identified.

12.4. Economy related to digital tools

Considering the usual ratios of firms, which approximately share the same size, we can note that firms having undergone the digital transformation have different characteristics:

– they impact more consumers;

– they employ few employees;

– they provide a large share of innovative global investment;

– their influence on the markets is larger;

– they pay low taxes thanks to tax optimization;

– they build the future;

– they build transnational programs and define standards. This is particularly the case of data structures (Big Data), artificial intelligence, robotics, technology or transhumanism (singularity).

As proposed by Westerman (2011), we have a new economic model, which calls into question the construction of value in organizations, as well as their boundaries.

Change in the relationship with the customer (customer experience)	Change in business processes	Change in business models
Understanding the customer	**Digitalization of processes**	**Digital transformation of organizations**
Segmentation analytical method Computerization of social networks	Improved performance (productivity) New functionalities	Increased products and services Transition from the physical to the digital Digital wrappers (digital network)
Growth of activities (turnover growth)	**Employee autonomy**	**New digital commerce**
Sales improved thanks to digital tools Predictive marketing Streamlining processes (integrated logistics)	Working anywhere, anytime (telecommuting) Larger, faster, more agile working community Sharing knowledge (and skills)	Increased products and services Redefinition of operational perimeters Redefinition of operational cycles
Point of contact with the customer	**Performance management**	**Globalization**
Customer service Consistency of communication channels Self-service	Operational transparency Data-driven decision-making	Integration of the firm in the world Redistribution of digital power Sharing of digital services
Digital tools		
Unification of the market data and processes		Management numerical integration
Analytical capacity	Numerical integration of production process elements	New solutions for delivering products and services

Table 12.3. *Challenges of digital transformation (source: Westerman, Cap Gemini, EY)*

12.5. The power of platforms

From bank to industry, from the public sector to trade and from BtoC to BtoB to BtoBtoC, every sector and model has been affected by the platform model.

12.5.1. *The concept of platform*

A first distinction should separate wide audience platforms, destined to the largest number, from business platforms, which are introduced as a kind of service, most often in the cloud, developed and controlled by one or more firms, and destined to employees, a group of customers or a group of clearly identified providers or partners.

This platform model intends to create new product–service couples. They can be built with the aim of becoming new sources of income for their developers. The goal can also be gaining efficiency or to become more attractive. In the latter case, the platform becomes a brand, which erases those from producers. Another goal is to improve customer experience and, consequently, to gain customer's loyalty. One of the major attitudes is to draw the maximum number of users to the platform. This requires an investment in the popularity of the platform and in the majority of cases, as well as the creation of innovative and multi-sectoral partnerships.

12.5.2. *The role of trust in platforms*

The concept of platform is somehow imperfect in the sense that it requires the applicants to "buy" to salespeople who are perfect strangers to them. Hence, the actors from collaborative economy use trust to convince potential buyers to carry out business on their platform. Thus, the best platform is not the most efficient one, but the one that inspires the highest trust. The construction of this trust, as we have already mentioned, involves rating. Trust in the platform takes the place of trust in the brand.

12.5.3. *The different types of platforms*

There are different types of platforms, which correspond to different visions of the new industry of society 5.0:

– factory 4.0: integrated supply chain, often presented as the German model;

– key-technology factory, based on a highly differentiating process;

– craft-industrial factory, in which tailor-made becomes industrialized.

– client drive factory, in which the customer operates the process;

– low-cost factory, including a large part of elements in Open Source and which are often self-propagating[4].

Carrefour, regarding the implementation of a digital platform, Carrefour.io, corresponds to the concept of factory 4.0, with an integrated logistics chain.

SNCF has set up a platform for the group's businesses. In this case, it is the implementation of skills related to key railway technologies, which become crucial.

Guest to Guest and AirBnB are examples of a new business based on a platform model with value-generating transversal partnerships These are client drive approaches, where customers are involved from the beginning of the process and make the first contact.

The data.gouv.fr platform at Etalab, an agency belonging to the French ministries, is one of the examples of achievements at the service of users and communities based on the platform model and Open Data. These are not customer-related platforms, but rather a low-cost approach based on Open Source.

4 3D printers are designed so that they can be built with the same 3D printer and a few additional elements.

12.5.4. *The State as platform*

> "In order to shape the public services of the 21ST century, the State has to "uberize" itself, in other words, it has to open itself to the multiplicity of users and entrepreneurs which could enable it to better reach general interest, and this, as quick as possible"[5].

Many countries started as scientists in the 1950s, by replicating initially paper procedures into the electronic form. "For a large group, e-Government was conceived as the simple computerization of silos forms, done administration by administration",[6] according to Laura Létourneau, the co-author of *Ubérisons l'État !* A digital logic of the State needs to be built, because digitalization cannot be envisioned as a solution, which will, on its own, make public services more efficient. The challenge is to break the silos. Then, it is necessary to adopt an external view, from the perspective of user needs, and transform their demands for public services thanks to digital technology. "Disintermediation" of the State services becomes a need, since the productivity, quality and efficiency of public services can be substantially increased and cost reduced. Tools should be put in place, together with the necessary forms of artificial intelligence, in order to understand and diagnose users' requests and implement solutions that will help avoid Kafkaesque administrative universes. One of the priority tasks is to build a transversal bond between the State's services and communities and the digital single counter that will then be built. In other words, governments must shift from remote procedures to platforms.

The challenges of the new services must eliminate the routine functioning of administrations and rebuild values of trust and cooperation. In other words, the State should also access the logic of "co" society. This platform-State will enhance collaboration between citizens and administrations. It should rely on territories and local initiatives. The State should become a watchman, which:

– properly disseminates new collaborative practices;

– very optimally uses new communication technologies;

– promotes social returns on the most effective investments.

---

5 Note from the FONDAPOL political studies foundation, July, 18th 2017.

6 Clément Bertholet, Laura Létourneau, *Ubérisons l'État !Avant que d'autres ne s'en chargent*, Armand Colin, Paris, 2017.

Trust in politics, which we know is shrinking, is linked to trust in the administration and especially in relations with users who themselves are reduced to “irritants”: saturated counters, abusive call centers, sclerotic school, Kafkaesque procedures, slow response from the administration. For this, it will be necessary to decompartmentalize institutions and public organizations and to encourage cooperation between them.

The challenge is to ensure that States are in a situation where they must undergo the assault from the outside and therefore the loss of their sovereignty. Therefore, a digital policy should promote the idea that the value of certain public services can be created out of its grip, while maintaining its sovereign field. For making sure that the State can guarantee general interest, it should adopt an analytical approach, making it possible to simultaneously:

– understand the issues of value creation and value use by resorting to other actors, particularly in the private sector and

– analyze the place of general interest to find whether there is a real productivity gain or a hidden loss of sovereignty.

In other countries, such platforms exist. Let us mention the Chinese health platform WeChat, which manages the relationship with hospitals. It makes it possible to make an appointment with the specialist of their choice, to obtain the results of health analyses, to make simple diagnostics remotely and also to pay for healthcare. The start-up Finamatic is revolutionizing the distribution of public aid. Public officials have more available time to assess entrepreneurs on special products, rather than to dedicate it to verify that administrative forms are correctly filled.

A concept of robust State still has to be built. This requires that the State and its administration:

– make sure that democratic and republican values are respected;

– defend rights;

– know how to build and improve their own applications;

– regulate the economy, the uses of the information and data.

The threat is a compromise between accessibility and merchandization, a capture through merchant services.

After having built thematic platforms, the State should become a kind of "meta" platform, which exercises power over these specialized platforms and over external platforms, under the sole condition of a win–win partnership.

There are several examples of merchandization. The National Library of France (BnF) accepts that free access works should be paying, since these cannot scan themselves, but is now requiring publishers also to provide digital versions. Genealogists' societies resell vital public data, but this is often done with additional services. Bob emploi signed a partnership agreement with Pôle emploi including a unique goal: to reduce frictional unemployment (also known as "mobility", transient and usually short-lasting unemployment) under a unique condition: not to make the unemployed to pay for that service. The Interdepartmental Management of the Digital and Information and Communication System of the State (DINSIC) has encouraged more than 20 digital "State start-up" services: the Open Data portal Data.gouv.fr, the Simplified Public Market (SPM), My aid gouv.fr, and Labonneboite.pole-Emploi.fr are some of the examples.

For those who support the State platform, such as the authors we have already quoted, civil society is ready to get involved in the co-construction of a public service, because "the State alone will not achieve any of its aims, nor even the SGMAP [General Secretariat for the Modernization of Public Action], nor the State secretariat for the digital nor the Ministry for Action and Public Accounts"[7] The transformation of the State is also that of the whole society and that of society 5.0 which is being born. In order to give birth to it, the strength of the multitude and that of millions of users will be more important than the political decisions from the public administration. It will be necessary to create "intrapreneurs" at the public service and to offer public servants a remuneration acting as an incentive for innovation, rather than offering salaries and promotions to the least intrepid public servants. These actions will involve failures, which will need to be analyzed, and commitment to risk will be valued.

---

7 Clément Bertholet and Laura Létourneau, *op. cit.*

The State will have to make a commitment to provide high-quality data. A great number of civil status files and electoral lists have poor quality. The success of the State platform will depend on the fact that tests are carried out, that the resistance to changes on the part of officials is reduced and that the platform model can spread. The challenge will be to create a new form of global leadership in areas that have remained unexplored by the American giants. This will be the challenge of society 5.0 when changes have not been made in time.

For some experts, a revolution of uses will intervene via new hardware, which are connected objects in the broad sense (this can refer to the simple sensor or to a complex system, in as much as a smartphone). At present, there are norms for their implementation. The example of care connected cabins is often mentioned. It is possible to get medical diagnosis through a video conference with the doctor and to carry out some tests, as well as blood analysis. The next step makes it possible to obtain the results almost immediately. These connected objects could be a partial solution to medical deserts.

12.5.5. *Platform as a service*

Platform as a service (PaaS) allow firms to facilitate the work of their developers by providing readily available work environments, which make it possible to make platforms.

As services platforms are a necessity for the actors of society 5.0, these technical solutions help organizations to quickly build these tools while differentiating themselves. Decision makers are more than ever eager to offer impeccable customer experience, hence the need to build solutions with method and quality. At the same time, technical solutions are in a perpetual technological change. Having the best developers is no longer enough in this competitive world. It is also necessary to have the best tools to save time in the face of competition. Developers can focus on quality, to added-value tasks, to adapt to the needs of the organization. PaaS thus becomes an essential tool, because they offer the organization a very rich, scalable, proven and tested technical base.

12.5.6. *Marketing platforms*

In marketing, firms integrate unique client relationship platforms or CDPs (customer data platforms). These are developments of the old tools for customer relationship management (CRM). At present, these platforms contain analytical, automation and optimization solutions in real time for e-marketing operations. Current developments are seeking a more comprehensive data collection. In general, the following three modules make up these platforms:

– one data collection platform, which collects online and offline data;

– one data decision platform, which analyzes the different marketing levers;

– one data management platform, which allows data management and activation to carry out specific actions.

Journal Technologies is one of these customer platform firms. Eulerian claimed more than 130 customers in 35 countries in 2017. The firm has 50 employees distributed in France, Spain and Canada. Technology claims more than 130 customers in 35 countries, including Voyages-sncf.com, Celio, SFR, Sarenza, Fnac, Galeries Lafayette and Warner Bros. In 2015, the firm achieved a turnover of EUR 5.7 million.

12.6. The limits of platforms

The situation of Uber is one of the examples of the limitations of platforms. Uber is a company that connects transport customers with drivers. Unlike the actors from other platforms, the advantage is that it is easy to find drivers due to the existence of unemployment in the large targeted neighborhoods. In this market, Uber transforms the fixed prices of taxi ride into a price that evolves according to supply and demand. Besides, the application has been promoted as quality service linking top-of-the-line drivers with customers. The new economic model proposed by Uber has two characteristics: it is not as original as communication suggests, reassures the financiers who provide money and appears as a disruptive activity, confronting a regulated profession. This firm has made itself a legend with the creation of the term "uberization".

Creation of Uber in 2009 by Travis Kalanick.

The start-up raised US $15 billion for a valuation of US $60 billion.

In 2017, Uber announced having employed more than 14,000 people and might have generated 5 billion trips.

The company is present in more than 600 major cities in the world, spread across 50 countries.

The company has a turnover of US $20 billion, generated US $6.5 billion in revenue and suffered US $2.8 billion in losses.

Box 12.1. *Uber's numbers*

The typical movement forward of firms using yield management as an economic model characterizes Uber. It has started delivering home meals in 72 cities, and it is implementing small logistics services and trying to design an autonomous vehicle. In cases like this, there is always a reduction in quality over time, and top-of-the-line drivers were finally recruited among inexperienced, suburban youngsters with little formal education.

The second problem with this type of firm is that the actors who sense approaching difficulties want to quit: this is particularly the case of shareholders. It is then necessary to address this situation before it is too late. Among the big players who might have invested in the firm are the founder of Amazon, Jeff Bezos, BlackRock, Sovereign Fund of Saudi Arabia, Goldman Sachs, Google, Morgan Stanley and Toyota.

12.7. Free economy

This form of economy was mainly developed with the advent of the Internet and the Web.

12.7.1. *The characteristics of free economy*

Free economy is a recent characteristic of economy. Traditionally, the following three types of free services should be differentiated:

– free as opposed to advertising, which is, for example, the case of television, some newspapers or some websites;

– free against data, an approach that certain platforms adore;

– a mix of the two.

12.7.2. *The example of the "free" newspaper market*

Metro, Direct Soir and 20 minutes are the best-known examples in France. Metro is present in 16 countries and globally has 15 million readers, knowing that a threshold of 10 million readers is enough to achieve financial equilibrium. Initially created in Canada under the name "24 hours", the regional newspaper Ouest-France is a 50% shareholder of "20 minutes". This journal is present in different cities like Lille, Bordeaux, Lyon and Marseille. Lille Plus is an example of a free newspaper created by the regional press. All paying newspapers are beginning to create free papers in order to avoid rivalry from precedent ones.

The way in which a free newspaper works is similar to the approach of a low-cost firm. The Ville Plus newspaper, to which Lille Plus belongs, is characterized by a small journalist workforce. Thus, the example of Lille's free newspaper that we are analyzing is made up of nine people in total: three journalists, four commercial employees and one secretary. The three journalists from Lille use the resources of nine of their colleagues located in Lyon. In addition, the newspaper has outsourced distribution.

12.8. The fight against large firms

The fight against large firms is in vogue once again due to the power of Google, Amazon, Facebook and Apple (GAFA).

This debate is old. The accumulation of capital gives big firms growing power. Our democratic societies are trying to limit the counter-force of politics. In the United States, the Sherman Antitrust Act of 1890 enacted the first rules against monopolies.

During the crisis of the 1930s, big firms were denounced as monsters, as "Frankenstein, Incorporated"[8]. After the war, multinational corporations were seen as the sword arm of Western policies in African countries and Western countries were perceived as colonialists and imperialists.

For some, these new multinational firms related to the digital are:

– unidentified political objects (UPOs);

– new forms of State[9].

In terms of society, its firms are presented in two ways:

– as avatars of autonomous political powers;

– as firms at the service of our well-being and facilitators of everyday life thanks to the innovations they bring.

Facebook is resolutely the first case. In a letter published on the network on February 16th, 2017, its owner, Mark Zuckerberg, wrote:

> "Our next goal will be to develop a social infrastructure for our community, to support it, to keep our security, to inform, to favor civic engagement and for the inclusion of all."

The problem that arises then is how to implement a type of control associated with the possibility of circumscribing the influence of these firms on political life and on democracy. This control is necessarily bound to data and processes associated with these data, such as data visualization.

12.9. The notion of data visualization

With the volume of data collected by society 5.0, data visualization tools are becoming increasingly necessary. These tools are becoming genuine platforms built in a specific way. Data visualization is a part of the construction of digital twins.

---

8 This is the title of the book by American lawyer Isaac Maurice Wormser (1887–1955), published in 1931.

9 This is the thesis of the Danish Minister of Foreign Affairs, Anders Samuelsen. On January, 26th 2017, he announced the creation of a "digital ambassador" position in Silicon Valley.

Data visualization tools may adopt different forms:

– available extensions for traditional tools (Excel, OpenOffice);

– opportunities offered by online office suites;

– simple online applications and services, generally offered by start-ups;

– JavaScript libraries to be integrated into existing applications;

– dashboard generators;

– totally integrated platforms with matching destination (table, Power BI, Qlik);

– visualization tools by means of diagrams and graphs.

Data visualization became light standby platforms, capable of extracting all freely accessible information via the Internet and of helping in the construction of knowledge. These platforms are also involved in the development of collaborative intelligence cells, for (re)defining the watch strategic objectives of firms and for product sourcing. The analysis of information is often facilitated through data visualization, to which we can add innovation, organization and animation methods.

12.10. Technology creating new resources

The smartphone, descendant of the mobile phone, is a typical example of the creative technologies of new resources. The book of Allard reminds us of an essential fact. The economic impact of the "mobile phone" as an object is non-negligible; it makes it possible to reduce the cost of production of motion picture films in the same way as a camcorder, webcam and so on. The volume of shareable production is also an issue of economic development, particularly of platforms and servers on which it is located.

Conclusion

How Can We Channel This Revolution?

Society 5.0 is inevitable, and the impact of the economic actors involved and the uncontrollable influence of the technologies on which it is built are undeniably two strong points. Unfortunately, we have to reckon that forecasting is not applicable, because society 5.0 means disruption. Therefore, the same applies to forecasting tools.

Can we still rely on an artistic world, which, like the writings of Jules Verne, could offer us some ideas about the future? Certainly, yes, we can! Therefore, the question lingers, or rather, a new one appears: how can we spot the artists at the confines of conventional and digital arts that could help us shed some light on the future?

Bibliography

AGAMBEN G., *Qu'est-ce qu'un dispositif?*, Rivages et Payot, Paris, 2007.

AGAMBEN G., *Qu'est-ce que le contemporain?*, Rivages et Payot, Paris, 2008.

AGARWAL R., PRASAD J.A., "A conceptual and operational definition of personal innovativeness in the domain of information technology", *Information Systems Research*, vol. 9, no. 2, pp. 204–215, June 1998.

AJZEN I., "The theory of planned behavior", *Organizational Behavior and Human Decision Processes*, vol. 50, no. 2, pp. 179–211, 1991.

AKRICH M., "Comment décrire les objets techniques?", *Techniques et culture*, no. 9, pp. 49–64, 1987.

AKRICH M., "Les objets techniques et leurs utilisateurs, de la conception à l'action", *Raisons pratiques*, no. 4, pp. 35–57, 1993.

AKRICH M., "Les utilisateurs, acteurs de l'innovation", *Éducation permanente*, no. 134, pp. 79–89, 1998.

ALHAWARI S., NEHARI T.A., ALRYALAT H. *et al.*, "The impact of knowledge process about customer on the success of customer knowledge acquisition", *Communications of the IBIMA*, vol. 1, no. 3, pp. 27–33, 2008.

ALLARD L., CRETON L., ODIN R. (eds), *Téléphone mobile et création*, Armand Colin, Paris, 2014.

ALTER N., *L'Innovation ordinaire*, PUF, Paris, 2000.

ALTHUSSER L., *Pour Marx*, Maspero, Paris, 1965.

AMMI C., *Le Marketing ethnique*, Hermes-Lavoisier, Paris, 2005.

ANDERSON C., *La Longue Traîne: La nouvelle économie est là!*, Dunod, Paris, 2006.

ANSOFF H.I., *The New Corporate Strategy*, John Wiley & Sons, Hoboken, 1965.

APPADURAL A., *Modernity at Large, Cultural Dimensions of Globalisation*, University of Minnesota Press, Minneapolis, 1996.

ASSOULINE P., "Y a-t-il un bon usage de Wikipédia? Entretien", *Le Débat*, no. 148, pp. 31–38, 2008.

ATAMER T., CALORI R., *Diagnostic et décisions stratégiques*, 2nd edition, Dunod, Paris, 2003.

ATTALI J., *Une brève histoire de l'avenir*, Fayard, Paris, 2006.

ATTALI J., Rapport de la Commission pour la libération de la croissance française: 300 décisions pour changer la France, destiné à la présidence de la République, La Documentation française, Paris, 2008.

AUGE M., *Non lieux, introduction à une anthropologie de la surmodernité*, Le Seuil, Paris, 1992.

AUGE M., *Où est passé l'avenir?*, Panama, Paris, 2008.

AUSTIN J.L., LANE G., RECANATI F., *Quand dire, c'est faire*, Le Seuil, Paris, 1970.

BADOT O., COVA B., *Le Néo-marketing*, Éditions ESF, Paris, 1992.

BALTZ C., BOULLIER D., *Vidéotex*, Centre national d'études des télécommunications, Issy-les-Moulineaux, 1984.

BALTZ C., "Une culture pour la société de l'information? Position théorique, définition, enjeux", *Documentaliste*, vol. 35, no. 2, pp. 75–82, 2008.

BANATHY BELA H., "Systems design of education", *A Journey to Create the Future*, Educational Technology Publications, Englewood Cliffs, 1991.

BANATHY BELA H., *Designing Social Systems in a Changing World*, Plenum Press, New York, 1996.

BARDINI T., "Changement et réseaux sociotechniques, de l'inscription à l'affordance", *Réseaux*, no. 76, pp. 126–155, 1996.

BARNEY J., "Firm resources and sustained competitive advantage", *Journal of Management*, vol. 17, no. 1, pp. 99–120, 1991.

BAUMAN Z., *S'acheter une vie*, Jacqueline Chambon Éditeur, Paris, 2007.

BECKETT S., *En attendant Godot*, Éditions de Minuit, Paris, 1952.

BENVENISTE E., *Problèmes de linguistique générale*, Gallimard, Paris, 1968.

BERELSON B., *Content Analysis in Communication Research*, Free Press, New York, 1952.

BERTHOLET C., LETOURNEAU L., *Ubérisons l'État! Avant que d'autres ne s'en chargent*, Armand Colin, Paris, 2017.

BOLTANSKI L., CHIAPELLO E., *Le Nouvel Esprit du capitalisme*, Gallimard, Paris, 1999.

BOLTANSKI L., *Rendre la réalité inacceptable. À propos de la Production de l'idéologie dominante*, Demopolis, Paris, 2008.

BOULLIER D., "Usages du Vidéotex et utopie techniciste", *Réseaux*, vol. 2, no. 6, pp. 21–36, 1984.

BOURDIEU P., *La Distinction. Critique sociale du jugement*, Éditions de Minuit, Paris, 1979.

BOURDIEU P., *Ce que parler veut dire: l'économie des échanges linguistiques*, Fayard, Paris, 1982.

BOURDIEU P., *Méditation pascalienne*, Le Seuil, Paris, 2003.

BRAUDEL P., *L'Identité de la France*, vol. 3, Arthaud, Paris, 1986.

BRETON P., *L'Utopie de la communication*, La Découverte, Paris, 1997.

BRUNS A., *Blogs, Wikipedia, Second Life and Beyond. From Production to Produsage*, Peter Lang Editor, New York, 2008.

CADIX A., Pour une politique nationale de design, essay submitted to the French Minister of Productive Recovery and Minister of Culture and Communication, October 2013.

CAHEN P., *Signaux faibles, mode d'emploi*, Eyrolles, Paris, 2010.

CAHEN P., *Le Marketing de l'incertain. Méthode agile de prospective par les signaux faibles et les scénarios dynamiques*, Kawa, Paris, 2011.

CAHEN P., *50 réponses aux questions que vous n'osez même pas poser*, Kawa, Paris, 2012.

CAHEN P., *Les Secrets de la prospective par les signaux faibles*, Kawa, Paris, 2013.

CAILLE A., *Critique de la raison utilitaire*, La Découverte, Paris, 2003.

CAILLE A., *Don, intérêt et désintéressement: Bourdieu, Mauss, Platon et quelques autres*, édition revue et augmentée, La Découverte, Paris, 2005a.

CAILLE A., *Dé-penser l'économique. Contre le fatalisme*, La Découverte, Paris, 2005b.

CAMUS R., *La Grande Déculturation*, Fayard, Paris, 2008.

CARON A.H., CARONIA L., *Culture mobile, les nouvelles pratiques de communication*, Presses de l'Université de Montréal, Montreal, 2005.

CARON F., *Les Voies de l'innovation, les leçons de l'histoire*, Manucius, Paris, 2011.

CARTON D., *La Démocratie Internet*, Le Seuil, Paris, 2010.

CASTELLS M., *La Galaxie Internet*, Fayard, Paris, 2001.

CASTORIADIS C., *L'Institution imaginaire de la société*, Le Seuil, Paris, 1975.

CATHELAT B., *Socio-styles système: " les styles de vie", théorie, méthode, applications*, Éditions d'Organisation, Paris, 1990.

CERTEAU (DE) M., *L'Invention du quotidien. Arts de faire*, 10/18, Paris, 1980.

CHAMBAT P., "Usages des TIC: évolution des problématiques", *Technologies de l'information et société*, vol. 6, no. 3, pp. 249–270, 1994.

CHARDEL P.A., ROCKHILL G. (eds), *Technologies de contrôle dans la mondialisation: enjeux politiques, éthiques et esthétiques*, Éditions Kimé, Paris, 2009.

CHARON D., *Introduction à la communication*, Presse de l'Université de Québec, Quebec, 1989.

CHAZAL G., *L'Ordre humain ou le déni de nature*, Éditions Champ Vallon, Seyssel, 2006.

CLÉMENT E., *La Philosophie de A à Z*, Hatier, Paris, 2004.

COLLIN T., *Individu et communauté, une crise sans issue?*, Mame, Paris, 2007.

COMTE A., *Cours de philosophie*, Hermann, Paris, 1998.

COMTE A., *Catéchisme positiviste*, Garnier-Flammarion, Paris, 1966.

CONVERSE P.D., *Essentials of Distribution*, Prentice Hall, New York, 1936.

CROUZET T., *Le Cinquième Pouvoir*, Bourin Éditeur, Paris, 2007.

DAVIS F.D., "Perceived usefulness, perceived ease of use, and user acceptance of information technology", *MIS Quarterly*, vol. 13, no. 3, pp. 319–340, 1989.

DAVIS F.D., BAGOZZI R.P., WARSHAW P.R., "User Acceptance of Computer Technology: A Comparison of Two Theoretical Models", *Management Science*, vol. 35, pp. 982–1003, 1989.

DEBORD G., *La Société du spectacle*, Buchet/Chastel, Paris, 1967.

DEMOULE J.P., *Les Dix Millénaires oubliés qui ont fait l'histoire*, Fayard, Paris, 2017.

DESJEUX D., "Usages et enjeux du SMS en Chine, en France et en Pologne", *Consommation & Sociétés*, available at: http://www.argonautes.fr/sections.php?op=viewarticle&artid=353, 2005.

DESREUMAUX A., LECOCQ X., WARNIER V., *Stratégie*, Pearson Éducation France, Paris, 2006.

DETRIE J.P. (ed.), *Stratégor: Politique générale de l'entreprise*, 3rd edition, Dunod, Paris, 2000.

DICHTER E., *Handbook of Consumer Motivations: the Psychology of the World of Objects*, McGraw-Hill, New York, 1964.

DOEHLEMANN M., *Langeweile? Deutung eines verbreiteten Phänomens*, Suhrkamp Verlag, Berlin, 1991.

DOKIC J., *Qu'est-ce que la perception?*, Vrin, Paris, 2009.

DREYFUSS H., *Symbol Sourcebook: An Authoritative Guide to International Graphic Symbols*, John Wiley & Sons, New York, 1984.

DUBY G., *L'Histoire continue*, Éditions Odile Jacob, Paris, 1991.

DUDEZERT A., *La Connaissance dans les entreprises*, La Découverte, Paris, 2013.

DUJARIER M.A., *L'Idéal au travail*, PUF, Paris, 2006.

DUJARIER M.A., *Le Travail du consommateur: de MacDo à eBay. Comment nous coproduisons ce que nous achetons*, La Découverte, Paris, 2008.

DURKHEIM E., *De la division du travail social*, PUF, Paris, 2007.

ECO H., *La Guerre du faux*, Grasset, Paris, 1985.

EIGLIER P., LANGEARD E., *Servuction. Le marketing des services*, 7th edition, Ediscience, Paris, 2000.

FAYOL H., *Administration industrielle et générale*, Dunod, Paris, 1918.

FEBVRE L., "De 1892 à 1933, examen de conscience d'une histoire et d'un historien (1934)", text republished in *Vivre l'histoire*, Laffont, Paris, 2009.

FERRARIS M., *T'es Où? Ontologie du téléphone mobile*, Albin Michel, Paris, 2006.

FISHBEIN B., MCGARRY L.S., DILLON P.S., *Leasing: A Step Toward Producer Responsibility*, INFORM, New York, 2000.

FISHBEIN M., AJZEN I., *Belief, Attitude, Intention, and Behavior: An Introduction to Theory and Research, Reading*, Addison-Wesley, Boston, 1975.

FISHER H., *Le Choc du numérique*, VLB Éditeurs, Montreal, 2001.

FLICHY P., *L'Innovation technique*, La Découverte, Paris, 1995.

FLICHY P., *Le Sacre de l'amateur*, Le Seuil, Paris, 2010.

FLUSSER V., *Choses et non-choses, esquisses phénoménologiques*, Éditions Jacqueline Chambon, Arles, 1996.

FLUSSER V., *Petite philosophie du design*, Éditions Circé, Belfort, 2002.

FOMBRUN C.G., *Reputation: Realizing Value from the Corporate Image*, Harvard Business School Press, Cambridge, 1996.

FORAY D., *L'Économie de la connaissance*, La Découverte, Paris, 2000.

FORAY D., ZIMMERMANN J.B., "L'économie du logiciel libre. Organisation coopérative et incitation à l'innovation", *Revue économique*, vol. 52, pp. 77–93, October 2001.

FORAY D., *Economics of Knowledge*, MIT Press, Cambridge, 2004.

FOUCAULT M., DEFERT D., EWALD F., *Dits et Écrits, 1954–1988*, vol. 3, Gallimard, Paris, 1994.

FREUD S., *Civilisation and its Discontents, Standard Edition of the Complete Psychological Works of Sigmund Freud*, Hoghart Press, London, 1973.

FRIEDL F., OTT N., STEIN B., *Typo. Quand, qui, comment. When, who, how. Wann, wer, wie*, Könemann, Cologne, 1998.

FUKUYAMA F., *La Fin de l'histoire et le dernier homme*, Flammarion, Paris, 1992.

GEERTZ C., *The Interpretation Of Cultures*, Basic Books Classics, New York, 1973.

GENEL K., "Le biopouvoir chez Foucault et Agamben", *Methodos*, vol. 4, 2004.

GENSOLLEN M., "La création de valeur sur Internet", *Réseaux*, vol. 97, pp. 15–76, 1999.

GERGEN K.J., "The challenge of absent presence", in KATZ J.E., AAKHUS M. (eds), *Perpetual Contact*, Cambridge University Press, Cambridge, 2002.

GIDDENS A., *Modernity and Self Identity, Self and Society in the Late Modern Age*, Stanford University Press, Stanford, 1991.

GIDDENS A., "L'émergence de la politique de vie", *Écologie Politique*, nos 11–12, pp. 137–158, 1994.

GILLE B., *Histoire des techniques*, Gallimard, Paris, 1978.

GILLILAND-SWETLAND A.J., "Setting the stage", in *Introduction to Metadata*, available at: https://www.getty.edu/publications/intrometadata/setting-the-stage, 2006.

GIRARD R., *Mensonge romantique et vérité romanesque*, Hachette, Paris, 2003.

GODARD F., *La Ville en mouvement*, Gallimard, Paris, 2001.

GODELIER M., *L'Énigme du don*, Fayard, Paris, 1997.

GONORD A., MENRATH J., *Mobile Attitudes: ce que les portables ont changé dans nos vies*, Hachette, Paris, 2005.

GRAMACCIA G., *Les Actes de langage dans les organisations*, L'Harmattan, Paris, 2001.

GRAMACCIA G., FAURE B., "La pragmatique des chiffres dans les organisations: de l'acte de langage à l'acte de calcul", *Études de Communication*, no. 29, pp. 25–37, 2007.

GRAMACCIA G., "Les actes de langage dans les organisations par projet", in BONNEVILLE L., GROSJEAN S. (eds), *Repenser la communication dans les organisations*, L'Harmattan, Paris, 2008.

GRAMACCIA G., "Qualité, projet, numérique: trois variations symboliques de l'efficacité gestionnaire", *Revue MEI*, no. 29, pp. 55–67, 2009.

GRAS A., *Les Macros-sytèmes techniques*, PUF, Paris, 1997.

GRÖNROOS C., "A service quality model and its marketing implications", *European Journal of Marketing*, vol. 18, no. 4, pp. 37–44, 1984.

GRÖNROOS C., "Toward a third phase in service quality research: challenges and future directions", *Advances in Services Marketing and Management*, vol. 2, pp. 49–64, 1993.

GRÖNROOS C., HEINONEN E., ISONIEMI K., "The net offer model: a case of example from the virtual maketspace", *Journal of Management History*, vol. 38, no. 4, p. 243, 2000.

GROSSEN M., PERRET-CLERMONT A.N., *L'Espace thérapeutique. Cadres et contextes*, Delachaux et Niestlé, Neuchâtel-Paris, 1992.

HABERMAS J., *L'Espace public*, Payot, Paris, 1978.

HAFFNER K., LYON M., "Le temps du mail. Esprit instantané et oral médiat", *Sociologie et Sociétés*, vol. 23, no. 2, pp. 154–171, 1999.

HARDT M., NEGRI A., *Multitude, Guerre et démocratie à l'âge de l'Empire*, La Découverte, Paris, 2004.

HEER J., Profiles as Conversation: Networked Identity Performance on Friendster, available at: http://www.danah.org/papers/HICSS2006.pdf, 2006.

HENDERSON V., *La Nature des soins infirmiers*, InterÉditions, Paris, France, 1994.

HERMALIN B., KATZ M.L., "Your network or mine? The economics of routing rules", *The RAND Journal of Economics*, vol. 37, no. 3, pp. 692–719, 2006.

HERT P., Jeux, écritures, espaces d'énonciations. Contribution à une étude anthropologique de l'usage d'Internet en milieu scientifique, Information and Communication Science PhD thesis, Université Strasbourg I, 1998.

HINCHCLIFFE D., KIM P., *Social Business by Design*, John Wiley & Sons, London, 2012.

HIPPEL (VON) E., *Democratizing Innovation*, MIT Press, Cambridge, 2005.

HOFSTEDE G., *Culture's Consequences: International Difference in Work-related Values*, Sage, Beverly Hills, 1980.

HOFSTEDE G., "The cultural relativity of organizational practices and theories", *Journal of International Business Studies*, vol. 14, no. 2, pp. 75–89, 1983.

HUNTINGTON S., "The clash of civilizations?", *Foreign Affairs, Council on Foreign Relations*, vol. 72, no. 3, pp. 22–49, 1993.

HUNTINGTON S., *Qui sommes-nous? Identité nationale et choc des cultures*, Odile Jacob, Paris, 2004.

HUNTINGTON S., *Revue Islamica*, available at: http://islamica magazine.com/?p=566, 2007a.

HUNTINGTON S., *Le Choc des civilisations*, Odile Jacob, Paris, 2007b.

HYMES D., "The ethnography of speaking", in GLADWIN T., STURTEVANT W.C. (eds), *Anthropology and Human Behavior*, The Anthropology Society of Washington, Washington, 1962.

HYMES D.H., GUMPERZ J.J. (eds), *Directions in Sociolinguistics*, Holt, Rinehart and Winston, New York, 1972.

HYMES D.H., FOUGHT J., *American Structuralism*, Mouton & Cie, La Haye, 1981.

HYPPOLITE J., *Introduction à la philosophie de l'histoire de Hegel*, Le Seuil, Paris, 1983.

IFRAH G., *Histoire universelle des chiffres*, Robert Laffont, Paris, 1994.

IHRY M.M., *Mythomania sur le Net*, Éditions Kirographaires, Marseille, 2013.

IRIBARNE (D') P., *La Logique de l'honneur*, Le Seuil, Paris, 1989.

IRIBARNE (D') P., *Penser la diversité du monde*, Le Seuil, Paris, 2008.

JACQUARD A., *La Matière et la Vie*, Milan, Toulouse, 1995.

JACQUARD A., *Nouvelle petite philosophie*, Stock, Paris, 2005.

JAUREGUIBERRY F., *Les Branchés du portable*, PUF, Paris, 2003.

JAUREGUIBERRY F., PROULX S., *Usages et enjeux des technologies de communication*, Éditions Érès, Toulouse, 2011.

JOANNIS H., *De l'étude de motivation à la création publicitaire et à la promotion des ventes*, 4th edition, Dunod, Paris, 1981.

JONSHON G., SCHOLES K., FRERY F., *Stratégique*, 7th edition, Pearson Éducation France, Paris, 2005.

JOUËT J., "Pratiques de la communication et figures de la médiation", *Réseaux*, no. 60, pp. 99–120, 2000a.

JOUËT J., "Retour critique sur la sociologie des usages", *Réseaux*, vol. 18, no. 100, pp. 487–521, 2000b.

JOULIN N., *Les Coulisses des nouveaux produits*, Eyrolles, Paris, 2002.

KATZ M.L., SHAPIRO C., "Network Externalities, Competition and Compatibility", *American Economic Review*, vol. 75, no. 3, pp. 424–440, June 1985.

KERCKHOVE (DE) D., *Les Nerfs de la culture, être humain à l'heure des machines à penser*, Les presses de l'Université de Laval, St-Nicolas, 1998.

KERCKHOVE (DE) D., *L'Intelligence des réseaux*, Éditions Odile Jacob, Paris, 2000.

KLING R., "Social issues and impacts of computing: from arena to discipline, in human choice and computers", in MOWSHOWITZ A. (ed.), *Proceedings of the Second IFIP-TC9 Human Choice and Computers Conference (HCC-2), Baden, 4–8 June, 1979*, Elsevier, Amsterdam, 1980.

KLING R., "Social analyses of computing: theoretical perspectives in recent empirical research", *Computing Surveys*, vol. 12, no. 1, pp. 61–110, 1980.

KLING R., *Computerization and Controversy. Value Conflict and Social Choice Book*, 2nd edition, Academic Press, San Diego, 1996.

KLING R., IT and Organizational changes in Digital Economies. A socio-technical approach, US Department of Commerce, available at: http://www.ecommerce.gov/schedule.htm, 25–26 May 1999a.

KLING R., "What is social informatics and why does it matter?", *D-Lib Magazine*, vol. 5, no. 1, available at: http://www.dlib.org/dlib/january99/klin/ 01kling.html, January 1999b.

KLING R., *What is Social Informatics?*, available at: http://www.slis.indiana.edu/SI/ concepts.html, 2001a.

KLING R., "Social Informatics", in *Encyclopedia of LIS*, Kluwer Publishing, Alphen-sur-le-Rhin, available at: http://www.slis.indiana.edu/SI/ si2001.Html, 2001b.

KOESTLER A., *Le Cri d'Archimède. La découverte de l'Art et l'art de la Découverte*, Les Belles Lettres, Paris, 2011.

KONDRATIEFF N., *Les Grands Cycles de la conjoncture*, Economica, Paris, 1992.

KOTLER P., *Marketing Management*, Prentice Hall, Engelwood Cliffs, 1967.

KREMER-MARIETTI A., *Le Positivisme d'Auguste Comte*, L'Harmattan, Paris, 2006.

KREMER-MARIETTI A., *Réflexions sur les temps actuels*, Espace de Libertés, Brussels, 2008.

KUNDERA M., *L'Identité*, Gallimard, Paris, 1998.

LADWEIN R., *Les Études marketing*, Economica, Paris, 1996.

LADWEIN R., *Le Comportement du consommateur et de l'acheteur*, 2nd edition, Economica, Paris, 2003.

LAKHANI K., WOLF R., Why Hackers Do What They Do: Understanding Motivation and Effort, In Free/Open Source Software Projects, Document de travail, MIT Sloan School of Management, no. 4425-03, 2005.

LAROQUE P., *Les Classes sociales*, PUF, France, 1977.

LAPOINTE L., RIVARD S., "Learning from physicians resistance to CIS implementation", *CMAJ*, vol. 174, no. 11, pp. 1573–1584, 2006.

LAUBIER (DE) C., "Les Français cassent leur tirelire pour le numérique", *Journal La Tribune*, vol. 20, 2008.

LAULAN A.M., *La Résistance aux systèmes d'information*, Éditions Retz, Paris, 1985.

LAZARSFELD P., BERELSON B., GAUDET H., *The People's Choice: How the Voter Makes up his Mind in a Presidential Campaign*, Duell, Sloan and Pearce, New York, 1944.

LAZARSFELD P.F., "An episode in the history of social research: a memoir", in FLEMING D., BAILYN B. (eds), *The Intellectual Migration: Europe and America, 1930–1960*, pp. 270–337, Harvard University Press, Cambridge, 1969.

LEDERER A.L. *et al.*, "The technology acceptance model and the World Wide Web", *Decision Support Systems*, vol. 29, no. 2000, pp. 269–282, 2000.

LEFEBVRE H., *La Vie Quotidienne dans le monde moderne*, Gallimard, Paris, 1968.

LEHIKOINEN J., AALTONEN A., HUSSKONEN P. *et al.*, *Personal Content Experience*, John Wiley & Sons, Hoboken, 2007.

LEJEALLE C., "Les usages ludiques du téléphone portable: bricoler avec un artefact technique", in ODIN F., THUDEROZ C. (eds), *Des mondes bricolés*, Presses polytechniques et universitaires romandes, Lausanne, 2008a.

LEJEALLE C., *Le Jeu sur téléphone portable, usages et sociabilité*, L'Harmattan, Paris, 2008b.

LEJEALLE C., *La Télévision mobile, usages, contenus, nomadisme*, L'Harmattan, Paris, 2009.

LEMARCHAND Y., *Du dépérissement à l'amortissement, enquête sur l'histoire d'un concept et de sa traduction comptable*, Ouest Editions, Nantes, 1993.

LEMARCHAND Y., NIKITIN M., Vingt ans d'histoire de la comptabilité, Comptabilité Contrôle Audi, 1999.

LE MOIGNE J.L. (eds), *Les Nouvelles Sciences. Comprendre les sciences de l'artificiel*, AFCET Éditions, Paris, 1984.

LE MOIGNE J.L., DEMAILLY A. (eds), *Sciences de l'intelligence, sciences de l'artificiel*, Presses Universitaires de Lyon, Lyon, 1986.

LEROI-GOURHAN A., *Le Geste et la Parole*, vol. 2, Albin Michel, Paris, 1964, 1965.

LEROY F., *Les Stratégies d'entreprise*, Dunod, Paris, 2002.

LERNER D., *The Passing of Traditional Society: Modernizing the Middle East*, The Free Press, New York, 1958.

LEVITT T., *The Marketing Imagination*, The Free Press, New York, 1983.

LEVY-BRUHL L., *La Mentalité primitive*, Éditions Alcan, Paris, 1922.

LEVY-BRUHL L., *L'Âme primitive*, Éditions Alcan, Paris, 1927a.

LEVY-BRUHL L., *La Morale et la Science des Mœurs*, Éditions Alcan, Paris, 1927b.

LIAUDET J.C., *Le Complexe d'Ubu ou la névrose libérale*, Fayard, Paris, 2004.

LICOPPE C., "Sociabilité et technologie de la communication. Deux modalités d'entretien des liens personnels dans le contexte du déploiement des dispositifs de communication mobiles", *Réseaux*, nos 112–113, 2002.

LIPOVETSKY G., SERROY J., *La Culture-monde, réponse à une société désorientée*, Éditions Odile Jacob, Paris, 2008.

LITWIN G., STRINGER R., *Motivation and organizational climate*, Harvard University Press, Cambridge, 1968.

LIVINGSTONE S., "The changing nature of audiences: from the mass audience to the interactive media user", in VALDIVIA A. (ed.), *Companion to Media Studies*, pp. 337–359, Blackwell Publishing, Oxford, 2003.

LUKACS G., *1921–1922, Histoire et conscience de classe. Essais de dialectique marxiste*, Éditions de Minuit, Paris, 1960.

LUSSATO B., FRANCE-LANORD B., BOUHOT P., *La Micro-informatique: introduction aux systèmes répartis*, Éditions d'informatique, Paris, 1974.

MAILLARD (DE) J., *Le Rapport censuré. Critique non autorisée d'un monde déréglé*, Flammarion, Paris, 2004.

MAKOWER J., "The Clean Revolution: Technologies from the Leading Edge", *Global Business Network Worldview Meeting*, 2001.

MALLEIN P., TOUSAINT Y., "Technologies de l'information et de la communication, une approche sociologique pour la conception assistée par l'usage", *Communication et Stratégie*, no. 5, pp. 77–99, 1994a.

MALLEIN P., TOUSAINT Y., "L'intégration sociale des technologies d'information et de communication: une sociologie des usages", *Technologies de l'Information et Société*, vol. 6, no. 4, pp. 315–334, 1994b.

MARCELLI D., *C'est en disant non qu'on s'affirme*, Hachette, Paris, 2007.

MARCO L. (ed.), *Nouvelles avancées du management*, L'Harmattan, Paris, 2005.

MARCUSE H., *Vers la libération. Au-delà de l'homme unidimensionnel*, Éditions de Minuit, Paris, 1969.

MARIEKE G., TONKIN E., "Folk-sonomies, Tidying up Tags?", *Dlib Magazine*, vol. 12, no. 1, available at: http://www.dlib.org/dlib/january06/guy/01guy.html, 2004.

MARTIN BARBERO J., *Communication, Culture and Hegemony, from the Media to the Mediation*, Sage Publishing, London, 1993.

MASLOW A., *L'Accomplissement de soi*, Eyrolles, Paris, 2004.

MASUDA Y., *The information society as post-industrial society*, Institute for the Information Society, Tokyo, 1980.

MATELLART A., *La Globalisation de la surveillance. Aux origines de l'ordre sécuritaire*, La Découverte, Paris, 2008.

MCCOMBS M., SHAW D., "The agenda-setting function of mass media", *Public Opinion Quarterly*, vol. 36, no. 2, p. 176, 1972.

MCKENNA R., *Le Marketing selon McKenna*, InterÉditions, Paris, 1985.

MCQUAIL D., *Sociology of Mass Communication*, Penguin, Harmondsworth, 1972.

MELMAN C., *L'Homme sans gravité. Jouir à tout prix, Entretiens avec Jean-Pierre Lebrun*, Denoël, Paris, 2002.

METCALF S.G., *10th World Multi-Conference on Systemics, Cybernetics and Informatics*, Orlando, Florida, United States, 16–19 July 2006.

MINTZBERG H., AHLSTRAND B., LAMPEL J., *Safari en pays stratégie*, Village mondial, Paris, 1998.

MISSIKA J.L., *La Fin de la télévision*, Le Seuil, Paris, 2006.

MODLEY R., *Handbook of Pictorial Symbols: 3,250 Examples from International Sources*, Dover Publishing, Mineola, 1976.

MOEGLIN P., "Du mode d'existence des outils pour apprendre", available at: http://w3.u-grenoble3.fr/les_enjeux/2000/Moeglin/home.html, 1999.

MOLES A., *Art et ordinateur*, Casterman, Paris, 1971.

MOORE G.C., BENBASAT I., "Development of an instrument to measure the perceptions of adopting an information technology innovation", *Information Systems Research*, vol. 2, no. 3, pp. 192–222, 1991a.

MOORE G.C., BENSABAT I., "Development of an instrument to measure the perception of adopting and information technology innovation", *Information System Research*, vol. 2, 1991b.

MORIN E., "Au-delà de la globalisation et du développement, société monde ou empire monde?", *Revue du Mauss*, no. 20, 2002.

MOWSHOWITZ A., *The Conquest of Will: Information Processing in Human Affairs*, Addison-Wesley, Boston, 1976.

MOWSHOWITZ A., *Inside Information: Computers in Fiction*, Addison-Wesley, Boston, 1977.

MOWSHOWITZ A., "Human choice and computers", *Proceedings of the Second IFIP Conference on Human Choice and Computers, Baden, 4–8 June 1979*, North-Holland Publication, Amsterdam, 1980.

MOWSHOWITZ A., "On the social relations of computers", *Human Systems Management*, vol. 5, pp. 99–110, 1985.

MOWSHOWITZ A., "Social dimensions of office automation", in YOVITS M. (ed.), *Advances in Computers*, vol. 25, pp. 335–344, Academic Press, New York, 1986.

MOWSHOWITZ A., "Le meilleur des marchés: marchandises informationelles et changement de société", in MAYERE A. (ed.), *La Société Informationelle*, pp. 23–42, L'Harmattan, Paris, 1997.

MOWSHOWITZ A., *Toward a Theory of Societal Transformation Stimulated by Information Technology*, Quorum Books, New York, 2002.

MUSSO P., *La Religion industrielle. Monastère, manufacture, usine. Une généalogie de l'entreprise*, Fayard, Paris, 2017.

NANCY J.L., *Juste impossible*, Bayard, Paris, 2007.

NDUBINSI N., SINTI Q., "Consumer attitudes systems characteristics, in internet banking", *Adoption in Malaysia, Management Research News*, vol. 29, 2006.

PATERNOSTER A., *Le Philosophe et les sens*, Presses Universitaires de Grenoble, Grenoble, 2009.

PENROSE E., *The Theory of the Growth of the Firm*, John Wiley & Sons, Hoboken, 1959.

PEPIN C., *Ceci n'est pas un manuel de philosophie*, Flammarion, Paris, 2010.

PERRIAULT J., *La Logique de l'usage. Essai sur les machines à communiquer*, Flammarion, Paris, 1989.

PERROT P., LAVELLE L., *Introduction à l'ontologie*, Éditions Le Félin, Paris, 2008.

PERROUX F., *Aliénation et société industrielle*, Gallimard, Paris, 1970.

PEZET A., PEZET E., *La Société managériale. Essai sur les nanotechnologies de l'économique et du social*, Éditions La ville brûle, Paris, 2010.

POIRIER N., "Bio-pouvoir: un concept incohérent et dangereux", *Les Temps modernes*, no. 640, pp. 156–162, 2006.

POIRIER N., "Critique de la notion de bio-pouvoir", *Cahiers critiques de philosophie*, no. 6, pp. 105–114, 2008.

POLANYI K., *The Great Transformation*, Rinehart, New York, 1944.

PORTER M., *La Concurrence selon Porter*, Pearson Éducation France, Paris, 1998.

PRAHALAD C.K., HAMEL G., "The core competencies of the corporation", *Harvard Business Review*, vol. 68, no. 3, pp. 79–93, 1990.

RAWL J., *Théorie de la justice*, Le Seuil, Paris, 1987.

RIFKIN J., *L'Âge de l'accès, la nouvelle culture du capitalisme*, La Découverte, Paris, 2000.

RIFKIN J., *La Troisième Révolution industrielle. Comment le pouvoir latéral va transformer l'énergie, l'économie et le monde*, Les Liens qui libèrent, Paris, 2012.

RINDOVA A., FOMBRUN C.J., "Constructing competitive advantage: the role of firm-constituent interactions", *Strategic Management Journal*, vol. 20, pp. 691–710, 1999.

ROGERS E.M., *Communication Technology: The New Media in Society*, The Free Press, New York, 1986.

ROGERS E.M., *Diffusion of Innovation*, 4th edition, The Free Press, New York, 1995.

RONELL A., *Telephone Book: Technologie, schizophrénie et langue électrique*, Bayard, Paris, 2006.

ROQUEPLO P., *Entre savoir et décision, l'expertise scientifique*, Éditions INRA, Versailles, 1997.

ROSNAY (DE) J., *Le Cerveau Planétaire*, Éditions Olivier Orban, Paris, 1986.

ROSNAY (DE) J., *La Révolte du pronétariat*, Fayard, Paris, 2006.

ROSOVSKYKY H., "Samuel Huntington, 81, political scientist, scholar. One of the most influential political scientists of the last 50 year", available at: News.harvard.edu/gazette/story/2009/02/samuel-huntington-81-political-scientist-scholar/, 5 February 2009.

RUBY C., *Devenir contemporain? La couleur du temps au prisme de l'art*, Éditions Le Félin, Paris, 2007.

RUSSELL B., *Problèmes de philosophie*, Paris, Payot, 1989.

RYLE G., *La Notion d'esprit*, Paris, Payot, 1978–2005.

SAID E., *L'Orientalisme, L'Orient créé par l'Occident*, Le Seuil, Paris, 1980.

SALGUES B., *Les Telecoms mobiles, GSM, DCS*, Hermes Science, Paris, 1997.

SALGUES B., *Territoires et société de l'information en France*, Collection INT, Evry, 2003.

SALGUES B., *L'E-Santé et la télémédecine*, Hermes-Lavoisier, Paris, 2013.

SALGUES B., ALLARD L., CRETON L. *et al.* (eds), "Téléphone mobile et création", *Questions de communication*, no. 26, pp. 318–319, available at: http://www.cairn.info/revue-questions-de-communication-2014-2-page-318.htm, 2014a.

SALGUES B., BISTER L., *Le Marketing des objets technologiques et des services numériques*, ISTE Editions, London, 2014b.

SALGUES B., *Health Industrialization*, ISTE Press, London and Elsevier, Oxford, 2016.

SANKAI Y., "cybernics: Fusion of Human, Machine and Information Systems", in SANKAI Y., SUZUKI K., HASEGAWA Y. (eds), *cybernics*, Springer, Tokyo, 2014.

SARTRE J.-P., *L'Être et le Néant*, Gallimard, Paris, 1943.

SARTRE J.-P., *Situations philosophiques*, Gallimard, Paris, 1990.

SCHAEFFER P., *Machines à communiquer*, vol. 1: *Genèse des simulacres*, Le Seuil, Paris, 1970.

SCHAEFFER P., *Machines à communiquer*, vol. 2: *Pouvoir et communication*, Le Seuil, Paris, 1972.

SCHELLING T., *La Stratégie du conflit*, PUF, Paris, 1960.

SCHMITT O.H., "A thermionic trigger", *Journal of Scientific Instruments*, vol. 15, pp. 24–26, 1938.

SCHNAPPER D., *L'Esprit démocratique des lois*, Gallimard, Paris, 2014.

SCHUMPETER J., *Capitalisme, socialisme et démocratie*, Payot, Paris, 1951.

SEABRIGHT P., *The Company of Strangers: A Natural History of Economic Life*, Princeton University Press, Princeton, 2010.

SEARLE J., *Les Actes de langage*, Hermann, Paris, 1972.

SEMPRINI A., *Analyser la communication, comment analyser les images, les médias la publicité?*, L'Harmattan, Paris, 1996.

SEN A., *On Economic Inequality*, Norton, New York, 1973.

SHANNON C.E., WEAVER W., *A Mathematical Model of Communication. Urbana*, University of Illinois Press, Champaign, 1949.

SIMON H.A., *Models of Bounded Rationality: Behavioral Economics and Business Organization*, vol. 1 and 2, The MIT Press, Cambridge, 1982.

SIMONDON G., *Du mode d'existence des objets techniques*, Aubier, Paris, 1958.

SLATER D., *Consumer Culture and Modernity*, Polity Press, Cambridge, 1997.

SN TZU, *L'Art de la Guerre*, Flammarion, Paris, 2017.

STIEGLER B., *La Télécratie contre la démocratie. Lettre ouverte aux représentants politiques*, Flammarion, Paris, 2006.

STIEGLER B., *Prendre soin, de la jeunesse et des générations*, Flammarion, Paris, 2008.

STIEGLER B., *Pour une nouvelle critique de l'économie politique*, Éditions Galilée, Paris, 2009.

STORA M., *Les Écrans, ça rend accro*, Hachette, Paris, 2007.

STOURDZE Y., *Organisation, anti-organisation*, Mame, Paris, 1973.

STOURDZE Y., La transmission instantanée: un révélateur social et culturel, Document ronéoté, Polycopié de cours, Université de Paris Dauphine IRIS, 1978.

STOURDZE Y., *Pour une poignée d'électrons*, Fayard, Paris, 1987.

SUZUKI K., "Augmented human technology", in SANKAI Y., SUZUKI K., HASEGAWA Y. (eds), *cybernics*, Springer, Tokyo, available at: https://doi.org/10.1007/978-4-431-54159-2_7, 2014.

SVENDSEN L.F.H., *Petite philosophie de l'ennui*, Fayard, Paris, 2003.

TAN M., TEO T.S.H., "Factors influencing the adoption of internet banking", *Journal of the Association for Information Systems*, vol. 1, 2000.

TAVERNIER P., "La sociocratie, un mode de gouvernance participatif", *La tribune fonda*, no. 224, available at: http://fonda.asso.fr/La-sociocratie-un-mode-de.html, 2014.

TOFFLER A., *La Troisième vague: essai*, Le Grand livre du mois, Paris, 1980.

TÖNNIES F., *Communauté et société. Catégories fondamentales de la sociologie pure*, PUF, Paris, 1977.

TORNATZKY L.G., KLEIN K.J., "Innovation characteristics and innovation adoption implementation, a meat analysis of findings", *IEEE Transaction on Engineering Management*, vol. 29, 1982.

USUNIER J.C., *Commerce entre cultures: Une approche culturelle du marketing international*, PUF, Paris, 1992.

VANDER W.T., *Explaining and Showing Broad and Narrow Folksonomies*, available at: http://www.vanderwal.net, 2005.

VENKATESH V., "Determinants of perceived ease of use: integrating control, intrinsic motivation, and emotion into the technology acceptance model", *Information Systems Research*, vol. 11, no. 4, pp. 342–365, 2000.

VENKATESH V. *et al.*, "User acceptance of Information technology. Toward a Unified View", *MIS Quarterly*, vol. 27, no. 3, pp. 425–478, 2003.

VERGELY B., *Petite philosophie pour jours tristes*, Milan, Toulouse, 2003.

VEYNE P., *Comment on écrit l'histoire*, Le Seuil, Paris, 1971.

VIRILIO P., *L'Horizon négatif*, Éditions Galilée, Paris, 1984.

VOGE J., RIGAL R., *Les Hyperfréquences, circuits et propagation des ondes: en vue des applications au radar et aux télécommunications*, Eyrolles, Paris, 1963.

VOGE J., "Crise, information et communication, avenir relatif des réseaux mailés et pyramidaux", *Communication & Medicine*, vol. 11, no. 3, 1985.

VOGE J., *Le Complexe de Babel, crise ou maîtrise de l'information?*, Éditions Masson, Paris, 1997.

WEBSTER F.E. JR., *Industrial Marketing Strategy*, John Wiley & Sons, Hoboken, 1995.

WEICK K., *The Social Psychology of Organizing*, 2nd edition, Random House, New York, 1979.

WEICK K., *Sensemaking in Organizations*, Sage, Thousand Oaks, 1995.

WIENER N., *Cybernetics, or Control and Communication in the Animal and the Machine*, MIT Press, Boston, 1948.

WINNER L., "Do artefacts have politics?", *Daedalus*, vol. 109, no. 1, pp. 121–136, 1980.

WITTGENSTEIN L., *Tractatus logico-philosophicus*, Gallimard, Paris, 1993.

WOLTON D., *Penser la communication*, Flammarion, Paris, 1997.

WOLTON D., *Internet et après? Une théorie critique des nouveaux médias*, Flammarion, Paris, 1999.

WOLTON D., *L'Autre mondialisation*, Flammarion, Paris, 2003.

WOOLGAR S., "Configuring the use: the case of usability trials", in LAW J. (ed.), *A Sociology of Monsters: Essays on Power, Technology and Domination*, Rouledge Editor, London, 1991.

ZARIFIAN P., *L'Échelle du monde. Globalisation, altermondialisme, mondialité*, La Dispute, Paris, 2004.

Index

D, E, F

G, H, I

J, K, L

M, N, O

P, Q, R

S, T, U

V, W, X, Y

Other titles from

iSTE

in

Science, Society and New Technologies

2018

BARTHES Angela, CHAMPOLLION Pierre, ALPE Yves
Evolutions of the Complex Relationship Between Education and Territories (Education Set - Volume 1)

DUGUÉ Bernard
Time, Emergences and Communications
(Engineering, Energy and Architecture Set – Volume 4)

GEORGANTOPOULOU Christina G., GEORGANTOPOULOS George A.
Fluid Mechanics in Channel, Pipe and Aerodynamic Design Geometries 1
(Engineering, Energy and Architecture Set – Volume 2)

GEORGANTOPOULOU Christina G., GEORGANTOPOULOS George A.
Fluid Mechanics in Channel, Pipe and Aerodynamic Design Geometries 2
(Engineering, Energy and Architecture Set – Volume 3)

LELEU-MERVIEL Sylvie
Informational Tracking
(Traces Set – Volume 1)

2017

ANICHINI Giulia, CARRARO Flavia, GESLIN Philippe,
GUILLE-ESCURET Georges
Technicity vs Scientificity – Complementarities and Rivalries
(Social Interdisciplinarity Set – Volume 2)

DUGUÉ Bernard
Information and the World Stage – From Philosophy to Science,
the World of Forms and Communications
(Engineering, Energy and Architecture Set – Volume 1)

GESLIN Philippe
Inside Anthropotechnology – User and Culture Centered Experience
(Social Interdisciplinarity Set – Volume 1)

GORIA Stéphane
Methods and Tools for Creative Competitive Intelligence

KEMBELLEC Gérald, BROUDOUS EVELYNE
Reading and Writing Knowledge in Scientific Communities– Digital
Humanities and Knowledge Construction

MAESSCHALCK Marc
Reflexive Governance for Research and Innovative Knowledge
(Responsible Research and Innovation Set - Volume 6)

PARK Sejin, GUILLE-ESCURET Georges
Sociobiology vs Socioecology – Consequences of an Unraveling Debate
(Interdisciplinarity between Biological Sciences and Social Sciences Set -
Volume 1)

PELLÉ Sophie
Business, Innovation and Responsibility
(Responsible Research and Innovation Set - Volume 7)

2016

BRONNER Gérald
Belief and Misbelief Asymmetry on the Internet

EL FALLAH SEGHROUCHNI Amal, ISHIKAWA Fuyuki, HÉRAULT Laurent, TOKUDA Hideyuki
Enablers for Smart Cities

GIANNI Robert
Responsibility and Freedom
(Responsible Research and Innovation Set - Volume 2)

GRUNWALD Armin
The Hermeneutic Side of Responsible Research and Innovation
(Responsible Research and Innovation Set - Volume 5)

LAGRANA Fernando
E-mail and Behavioral Changes – Uses and Misuses of Electronic Communications

LENOIR Virgil Cristian
Ethical Efficiency – Responsibility and Contingency
(Responsible Research and Innovation Set - Volume 1)

MAESSCHALCK Marc
Reflexive Governance for Research and Innovative Knowledge
(Responsible Research and Innovation Set - Volume 6)

PELLÉ Sophie, REBER Bernard
From Ethical Review to Responsible Research and Innovation
(Responsible Research and Innovation Set - Volume 3)

REBER Bernard
Precautionary Principle, Pluralism and Deliberation – Sciences and Ethics
(Responsible Research and Innovation Set - Volume 4)

VENTRE Daniel
Information Warfare – 2nd edition

Printed and bound by CPI Group (UK) Ltd, Croydon, CR0 4YY

15/09/2026

14962349-0004